WAVELET BASICS

WAVELET BASICS

by

Y. T. Chan

Royal Military College of Canada

KLUWER ACADEMIC PUBLISHERS
Boston / Dordrecht / London

Distributors for North America:
Kluwer Academic Publishers
101 Philip Drive
Assinippi Park
Norwell, Massachusetts 02061 USA

Distributors for all other countries:
Kluwer Academic Publishers Group
Distribution Centre
Post Office Box 322
3300 AH Dordrecht, THE NETHERLANDS

Consulting Editor: Jonathan Allen, Massachusetts Institute of Technology

Library of Congress Cataloging-in-Publication Data

A C.I.P. Catalogue record for this book is available from the Library of Congress.

Contents

4. Current Topics   111

Preface

This book had its humble beginning one morning in April, 1993. I was then on my second sabbatical leave at the Chinese University of Hong Kong and having the time of my life. That morning, as I was leaving for the office as usual with my tennis racket and gym bag, May told me in a normal wifery manner that my court time lately had been fast approaching office time and that I should do something about the situation. My main excuse for the sabbatical was to learn about wavelets and I decided that morning the best way to learn is to write a book on it.

It was quite a struggle. There were no introductory level texts for engineers and it was not easy to learn from the literature. Since it is a new subject, its terminology is non-standard, concepts and proofs are hard to follow and were at times confusing. I wrote this book for beginners, aiming to help readers avoid many of my own difficulties. The terminology is familiar to engineers. Whenever possible, I give examples to illustrate new concepts, answer questions that I encountered when studying the subject, and give intuitive insight.

As it turned out, my six months at the Chinese University were rather fruitful. Dr. P. C. Ching, of the department of electronic engineering at CUHK, made the arrangement for the visit and deserves a sincere thank you. This book will not be possible without the extraordinary efforts of Dr. K. C. Ho, a research fellow here at the Royal Military College. He not only typed the entire manuscript, did all the figures, but also corrected my mistakes and suggested proofs. I owe him a great deal. As to May, who is responsible for my last one and half years of misery, I cannot accord her the customary gratitude of having been understanding, supportive and tolerant, etc.

Y. T. CHAN
Kingston, Ontario, Canada

WAVELET BASICS

Chapter 1

INTRODUCTION

1.1 Signal Transformation

The transformation of a function or signal $s(t)$ is a mathematical operation that results in a different representation of $s(t)$. The well known Fourier transform gives the spectrum of a signal while a two dimensional transformation of an image may aim to concentrate the image energy into a smaller region for compression purposes. A prism acts as a Fourier transformer by decomposing sunlight into its visual spectrum of different colours (frequencies). Hence a transform also reveals the composition of a signal in terms of the building blocks, or basis functions, of the transformed domain. In the Fourier domain, the building blocks are sinusoids. A signal has a unique representation in the Fourier domain as a continuous sum of sinusoids of different amplitudes, frequencies and phases. On the other hand, the simple Walsh transform has basis functions that are variable width pulse sequences of amplitudes ± 1, as shown in Figure 1.1, where it is assumed without loss of generality that $s(t)$ is of duration from $t = 0$ to $t = 1$.

The Fourier transform pair is

$$S(\omega) = \int_{-\infty}^{\infty} s(t) e^{-j\omega t} dt \qquad (1.1)$$

$$s(t) = \frac{1}{2\pi} \int_{-\infty}^{\infty} S(\omega) e^{j\omega t} d\omega \qquad (1.2)$$

The decomposition of $s(t)$ is via (1.1), the Fourier transform. It decomposes $s(t)$ into sinusoids of frequency ω, amplitude $|S(\omega)|$ and phase $\angle S(\omega)$. The inverse Fourier transform, (1.2), synthesizes $s(t)$ from the basis functions $e^{j\omega t}$, of complex amplitude $S(\omega)$. Another way to view (1.1) is that the weight $S(\omega)$ is the "amount" of $e^{j\omega t}$ that $s(t)$ contains. Hence the cross-correlation of $s(t)$ with $e^{-j\omega t}$ yields $S(\omega)$. A simple parallel of (1.1) is in the determination of the coefficients α and β of the basis vectors $e_1 = [1 \quad 0]^T$, $e_2 = [0 \quad 1]^T$ needed to synthesize a particular vector v :

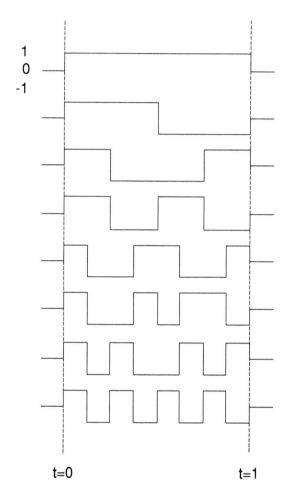

Figure 1.1 Some typical Walsh functions

$$v = \begin{bmatrix} a \\ b \end{bmatrix} = \alpha \begin{bmatrix} 1 \\ 0 \end{bmatrix} + \beta \begin{bmatrix} 0 \\ 1 \end{bmatrix} \tag{1.3}$$

To find α, take the inner product (cross-correlation) of v and e_1, i.e.

$$\alpha = <v, e_1> = a \tag{1.4}$$

and

$$\beta = <v, e_2> = b \qquad (1.5)$$

Or, the projections of v onto e_1 and e_2 give α and β respectively. Using simple basis functions, e.g. Walsh, will greatly simplify the transformation, or inner product, calculations. However, computation load is usually only one of several factors in choosing a particular transform. Others are a transform's properties and its suitability for a given application.

The reasons for transforming or decomposing a signal are numerous. The Laplace transform is a generalization of the Fourier transform and expresses a function $x(t)$ as a weighted, continuous sum of the basis function e^{st}. Thus

$$x(t) = \int_{-\infty}^{\infty} X(s) \, e^{st} \, ds \qquad (1.6)$$

where the weight $X(s)$ is the Laplace transform of $x(t)$ and s is a complex quantity called the complex frequency. As easily seen from (1.6), the equivalent operations in the Laplace domain of differentiation or integration of $x(t)$ in the time domain are multiplications of $X(s)$ by s or $1/s$. Thus taking the Laplace transform of a linear integro-differential equation will change it into an algebraic equation. This important result is the foundation of linear system analysis by the Laplace transform.

More recently, with the introduction of the Fast Fourier Transform (FFT), there is sometimes a speed gain in doing calculations in the frequency domain, for time domain operations such as convolution and correlation. Modern radar and sonar receivers invariably have an FFT front end and functions such as matched filtering and beam-forming are performed in the frequency domain. Signal transformations, by virtue of offering an alternate representation, often reveal key features of a signal that are difficult or impossible to discern in the original domain. The existence and locations of multiple periodicities, spectral and phase patterns, for example, are useful features in the frequency domain for detection and classification. Yet another important transform application is data compression. Let Φ be an $N \times N$ matrix whose elements are samples of an image. A transform operation gives

$$\Psi = W^T \Phi W \qquad (1.7)$$

A judicious choice of the transform matrix W could produce a transformed image matrix Ψ that is sparse and with most of its large magnitude elements concentrated in a small region of Ψ. This is the idea of decorrelation and energy compaction by transformation. Data compression is by transmitting only the elements in that small region. The receiving end recovers the data compressed image by inverse transformation. Typically only 15% of the elements of Ψ need be retained without an adverse effect on the image quality [1]. Examples of W are the Discrete cosine transform, Haar and Hadamard transforms [2].

1.2 Orthogonal Transforms

In Section 1.1, the example showing the calculations of the coefficients α and β conveniently chooses two basis vectors e_1 and e_2 which are orthogonal, i.e., $<e_1,e_2>=0$ and orthonormal, i.e. $<e_1,e_1>=<e_2,e_2>=1$. A set of vectors $\{v_i\}$, $i=1,2,\ldots,n$ is orthogonal if

$$<v_i,v_j>=c\,\delta_{ij} \tag{1.8}$$

where the Kronecker delta

$$\delta_{ij}=\begin{cases}0 & , & i\neq j \\ 1 & , & i=j\end{cases} \tag{1.9}$$

and c is a constant. It is orthonormal if $c=1$. When two vectors are orthogonal, they have no correlation or common components. The projection of one onto another is zero, and their inner product is zero. Decomposition of a vector into its components of orthonormal basis vectors is therefore rather simple. Let $\{v_i\}$ be a set of orthonormal vectors that spans the n-dimensional space, then any $n\times 1$ vector g is a linear combination of the v_i given by

$$g=\sum_{i=1}^{n}<g,v_i>v_i \tag{1.10}$$

If $\{ v_i \}$ is not orthonormal, g can still be expressed as a linear combination of $\{ v_i \}$, but the coefficients of v_i are no longer simple inner products of $<g, v_i>$. Let those coefficients be $h_1, h_2, \ldots h_n$,

$$g = V \begin{bmatrix} h_1 \\ h_2 \\ \vdots \\ h_n \end{bmatrix} \quad , \quad V = [\, v_1 \quad v_2 \quad \cdot \quad \cdot \quad \cdot \quad v_n \,] \tag{1.11}$$

and the operation $V^{-1} g$ is needed to find the coefficients.

In image coding, transformation of an image into its orthogonal components is decorrelation: the break up of an image into parts that have no similarities. Now because of their independence, the dropping from the transformed image of some orthogonal components will not affect the others. This important property of orthogonality is formally summarized by the projection theorem [3]. Let $\{ v_i \}$, $i = 1, 2, \ldots, n$ be a set of vectors that are orthonormal and span the n-dimensional space so that any vector g is

$$g = \sum_{i=1}^{n} <g, v_i> v_i \tag{1.12}$$

Let

$$\hat{g} = \sum_{i=1}^{l} <g, v_i> v_i \quad , \qquad l < n \tag{1.13}$$

Without loss of generality, (1.13) assumes that the vectors dropped are v_i, $i = l+1, l+2, \ldots, n$. The theorem states that $\hat{g}$ best approximates g in the least squares sense, i.e.

$$\| g - \hat{g} \|^2 \leq \| g - \tilde{g} \|^2 \tag{1.14}$$

where $\tilde{g}$ is any other approximation of g in terms of the reduced set of vectors v_i, $i = 1, 2, \ldots, l$. The proof is by writing

$$\tilde{g} = \sum_{i=1}^{l} \alpha_i v_i \quad , \qquad \alpha_i \text{ any constant} \tag{1.15}$$

and showing that $|g - \tilde{g}|^2$ is minimum when $\alpha_i = <g, v_i>$. The approximation error is

$$\varepsilon^2 = \| \sum_{i=1}^{n} <g, v_i> v_i - \sum_{i=1}^{l} \alpha_i v_i \|^2 \qquad (1.16)$$

On using the orthonormal properties of v_i, (1.16) becomes

$$\varepsilon^2 = \sum_{i=1}^{l} (<g, v_i> - \alpha_i)^2 + \sum_{i=l+1}^{n} |<g, v_i>|^2 \qquad (1.17)$$

which attains a minimum when $\tilde{g} = \hat{g}$. Moreover, the residual error, which is the second term in (1.17), decreases as l increases. To repeat, dropping any component of an orthonormal transformation results in a truncated representation that is still best in the least squares sense.

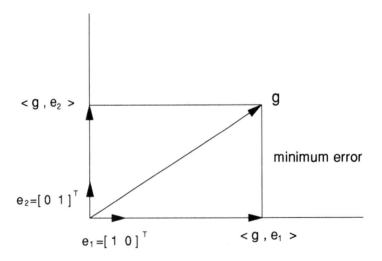

Figure 1.2 Projection of a vector

Geometrically, the vector g in Figure 1.2 has orthogonal components $<g, e_1>$ and $<g, e_2>$. If the e_2 component is dropped, the representation $\hat{g} = <g, e_1> e_1$ is still the least squares approximation to g because the shortest distance between g and e_1 is the projection of g on e_2.

For an expansion of g in the form of (1.10), it is easy to show that

$$\| g \|^2 = \sum_{i=1}^{n} |<g, v_i>|^2 \tag{1.18}$$

Thus the squared norm of the vector in the original domain equals the sum of the squares of the coefficients in the transform domain. This is equivalent to the Parseval's theorem for Fourier transform [3],

$$\int_{-\infty}^{\infty} s^2(t) \, dt = \frac{1}{2\pi} \int_{-\infty}^{\infty} |S(\omega)|^2 \, d\omega \tag{1.19}$$

which states that the energy in the time domain equals the energy in the frequency domain. Energy preserving transforms are called isometric transforms.

We next extend the principles of orthonormal expansion to functions. The set of functions $\{ f_i(t) \}$ are orthonormal over an interval t_1 and t_2 if

$$\frac{1}{t_2 - t_1} \int_{t_1}^{t_2} f_i(t) f_j^*(t) \, dt = \delta_{ij} \tag{1.20}$$

where * denotes the complex conjugate. For example, the set $\{ e^{jn\omega_o t} \}$, n an integer and ω_o a non-zero constant, is orthonormal over the interval $-T/2$ to $T/2$, for $T = 2\pi/\omega_o$. The least squares approximation property in vectors carries over to functions as well. Hence in the Fourier series representation of a periodic signal, the remaining sum, after a truncation of the series, is the least squares approximation of the signal in terms of the remaining terms.

In the representation of a signal by an orthonormal set $\{ f_i(t) \}$, an exact representation is possible only if the set is complete. A set is complete if there is no other function $h(t) \neq 0$, not in the set, such that

$$\int_{t_1}^{t_2} h(t) f_i^*(t) \, dt = 0 \quad , \qquad i = 1, 2, \ldots \tag{1.21}$$

If such a $h(t)$ exists, it is orthogonal to the set and should therefore be a member of the set, otherwise the set will not be complete. For vectors, a complete set means the set of basis vectors spans the vector space.

1.3 Frames in Vector Space

The decomposition of a vector into an orthonormal set of basis vectors is a simple inner product operation. Now it is desired to retain the same simple operation when the basis vectors are no longer orthonormal (or orthogonal). Note that basis vectors need not be orthonormal; they can even be linearly dependent and hence redundant. The only requirement is that they span the vector space so that any vector can be represented in terms of them. The theory of frames is a generalization of the orthonormal decomposition principle and gives a representation of an $m \times 1$ vector as

$$g = \sum_{i=1}^{n} <g, v_i> \tilde{v}_i \quad , \qquad n \geq m \tag{1.22}$$

This is similar to (1.10), except $\{ v_i \}$ is not necessarily orthonormal and since $n \geq m$, the v_i basis vectors can be linearly dependent. The $\{ \tilde{v}_i \}$ is called the dual of $\{ v_i \}$. What (1.22) states is that the simple inner product form is still valid in decomposition, but at the expense of introducing duals in the reconstruction.

The v_i are now called elements of a frame [4], and $\{ \tilde{v}_i \}$ the dual frame of $\{ v_i \}$. For simplicity, we assume v_i are unit vectors. A frame $\{ v_i \}$ is a set of vectors that satisfies, for any non-zero $m \times 1$ vector g,

$$A \| g \|^2 \leq \sum_{i=1}^{n} |<g, v_i>|^2 \leq B \| g \|^2 \quad , \qquad n \geq m \tag{1.23}$$

where A and B are constants, dependent on $\{ v_i \}$ only, called frame bounds, with $0 < A \leq B < \infty$. They are the highest lower bound and lowest upper bound. The lower bound guarantees that the set $\{ v_i \}$ spans the vector space, i.e., $\{v_i\}$ is a complete frame, otherwise $\sum_{i=1}^{n}|<g,v_i>|^2$ can become zero for some $\|g\|\neq 0$. If $\{v_i\}$ is a frame, then (1.22) holds. A frame is tight if $A = B$ and (1.22) holds with $\tilde{v}_i = v_i/A$. Further, if removal of a single element in a tight frame violates the lower bound of (1.23), i.e., the frame becomes incomplete, then the tight frame is an exact frame. Finally, elements of a tight, exact frame with $A = B = 1$ form an orthonormal basis and (1.22) holds with $\tilde{v}_i = v_i$.

In summary, the theory of frames provides the representation of a vector in terms of a set of basis vectors that are not necessarily orthonormal, nor linearly independent. The coefficients are still inner products of the vector with the basis vectors. Reconstruction, however, requires new basis vectors called duals. As long as $\{v_i\}$ obeys (1.23), any vector g can be synthesized according to (1.22). If $A = B$, then $\tilde{v}_i = v_i/A$ and if $\{v_i\}$ is tight, exact and $A = B = 1$, then $\tilde{v}_i = v_i$ and $\{v_i\}$ forms an orthonormal basis. The theory of frames is used later in the decomposition and reconstruction of a function by wavelets. We next give numerical examples, instead of proofs, to illustrate the above.

Example 1.1

Let

$$[v_1 \quad v_2 \quad v_3] = \begin{bmatrix} 1 & -\dfrac{1}{2} & -\dfrac{1}{2} \\ 0 & \dfrac{\sqrt{3}}{2} & -\dfrac{\sqrt{3}}{2} \end{bmatrix} \qquad (1.24)$$

then for any g,

$$\sum_{i=1}^{3}|<g,v_i>|^2 = \frac{3}{2}\|g\|^2 \qquad (1.25)$$

and it follows from (1.23) that $A = B = 3/2$, i.e. $\{v_i\}$ is a tight frame and $\tilde{v}_i = 2/3\, v_i$.

Example 1.2

Let

$$[v_1 \quad v_2 \quad v_3 \quad v_4] = \begin{bmatrix} 1 & 0 & -1 & 0 \\ 0 & 1 & 0 & -1 \end{bmatrix} \tag{1.26}$$

then for any g,

$$\sum_{i=1}^{4} |<g, v_i>|^2 = 2\|g\|^2 \tag{1.27}$$

and

$$g = \sum_{i=1}^{4} <g, v_i> \frac{v_i}{2} \tag{1.28}$$

Example 1.3

Let

$$[v_1 \quad v_2 \quad v_3] = \begin{bmatrix} 0 & -\dfrac{1}{2} & -\dfrac{1}{2} \\ 1 & \dfrac{\sqrt{3}}{2} & -\dfrac{\sqrt{3}}{2} \end{bmatrix} \tag{1.29}$$

and $g = [a \quad b]^T$ so that

$$\sum_{i=1}^{3} |<g, v_i>|^2 = \frac{1}{2}(a^2 + 5b^2) \tag{1.30}$$

and

$$\frac{1}{2}(a^2 + b^2) \le \sum_{i=1}^{3} |<g, v_i>|^2 \le \frac{5}{2}(a^2 + b^2) \tag{1.31}$$

Here $A = 1/2$, $B = 5/2$. The set $\{v_i\}$ is a frame but it is not tight. As will be shown in the next paragraph,

$$\tilde{v}_i = \begin{bmatrix} 2 & 0 \\ 0 & \dfrac{2}{5} \end{bmatrix} v_i \tag{1.32}$$

The preceding examples show that if $\{v_i\}$ is a tight frame, then $\tilde{v}_i = c\, v_i$, where c is a constant. To find the dual frame $\{\tilde{v}_i\}$ when $\{v_i\}$ is not tight, let

$$V_{m \times n} = [v_1 \quad v_2 \quad . \quad . \quad . \quad v_n] \quad ,$$

$$\tilde{V}_{m \times n} = [\tilde{v}_1 \quad \tilde{v}_2 \quad . \quad . \quad . \quad \tilde{v}_n] \tag{1.33}$$

Suppose $\{v_i\}$ obeys (1.23) so that it follows from (1.22) that

$$g = \tilde{V}\, V^T g \tag{1.34}$$

for any $m \times 1$ vector g. Thus $\tilde{V}$ must satisfy

$$\tilde{V}_{m \times n}\, V^T_{n \times m} = I_{m \times m} \tag{1.35}$$

where I is an identity matrix. In (1.35), there are mn unknowns in $\tilde{V}$ and m^2 $(n \geq m)$ available equations. The solution for $\tilde{V}$ is therefore not unique. The pseudo-inverse [5] solution is

$$\tilde{V} = (V\, V^T)^{-1} V \tag{1.36}$$

The inverse in (1.36) exists because of (1.23), which ensures that $\{v_i\}$ is a complete frame. Direct substitution of (1.36) into (1.35) of course yields the identity matrix. However, (1.36) is only the minimum norm solution. Any $\overline{V}$, with $\|\overline{V}\| \neq 0$ and $\overline{V}\, V^T = 0$ will give rise to another solution $\tilde{V} + \overline{V}$. Returning to Example 1.1, where we had $\tilde{V} = 2/3\, V$, it is easy to verify that $\tilde{V} = 2/3\, V + M$, where M is a matrix of unity, will also satisfy (1.22). Hence the dual frame is not unique, as to be expected, unless we impose the minimum norm constraint.

A proof is given next for the result that $\{v_i\}$ is an orthonormal set if and only if $A = B = 1$ in (1.23) and $\{v_i\}$ is an exact frame. Suppose $\{v_i\}$ is an orthonormal set, then for any $m \times 1$ vector g,

$$\sum_{i=1}^{m} |<g, v_i>|^2 = g^T V\, V^T g = \|g\|^2 \tag{1.37}$$

and it follows from (1.23) that $A = B = 1$. Next, suppose $A = B = 1$. Let

$$F = V V^T \tag{1.38}$$

which is a symmetric positive definite matrix because V has full rank. Then there exists a unitary matrix P so that

$$P^T F P = \Lambda \tag{1.39}$$

is a diagonal matrix with elements equal to the eigenvalues of F [6]. Hence

$$\sum_{i=1}^{m} |<g, v_i>|^2 = g^T P^T \Lambda P g \tag{1.40}$$

Since P is unitary, $\|P g\| = \|g\|$ so that

$$\lambda_{min} \|g\|^2 \le \sum_{i=1}^{m} |<g, v_i>|^2 \le \lambda_{max} \|g\|^2 \tag{1.41}$$

where λ_{min} and λ_{max} are the minimum and maximum eigenvalues (all positive) of F. If $\lambda_{min} = \lambda_{max} = \overline{\lambda}$, then F must be diagonal with all elements equal to $\overline{\lambda}$ and $\tilde{v}_i = v_i / \overline{\lambda}$. But if $\overline{\lambda} = 1$, i.e., $A = B = 1$, then $\tilde{v}_i = v_i$ and $\{v_i\}$ is orthonormal.

Resolution of the identity, introduced in Section 2.2, states that if a transformation is invertible, then the signal energy in the original domain, must be equal to, within a constant, the signal energy in the transform domain. An example of this identity is the Parseval's theorem. Thus a reconstruction of the signal, in terms of the basis functions, is feasible only if energy preservation holds within a constant. Frames in general do not satisfy resolution of the identity. Hence there is the need to introduce duals in the reconstruction. In the case of vector transformation, it is easy to verify that if $\{v_i\}$ is an orthonormal set, then

$$g = \sum_{i} <g, v_i> v_i \tag{1.42}$$

and

$$\|g\|^2 = \sum_i |<g, v_i>|^2 \tag{1.43}$$

i.e., resolution of the identity holds. For Examples 1.1 and 1.2, $A = B \neq 1$ so that

$$g = \frac{1}{A} \sum_i <g, v_i> v_i \tag{1.44}$$

and

$$\|g\|^2 = \frac{1}{A} \sum_i |<g, v_i>|^2 \tag{1.45}$$

For Example 1.3, $A \neq B$,

$$g = \sum_i <g, v_i> \tilde{v}_i \tag{1.46}$$

and

$$\|g\|^2 \neq c \sum_i |<g, v_i>|^2 \tag{1.47}$$

for any constant c. When a transform violates resolution of the identity, duals are needed for reconstruction. Finally, since

$$\sum_i <g, v_i> \tilde{v}_i = \tilde{V} V^T g \tag{1.48}$$

we can easily conclude by using (1.35) that

$$g = \sum_i <g, v_i> \tilde{v}_i = \sum_i <g, \tilde{v}_i> v_i \tag{1.49}$$

1.4 Time-Frequency Analysis

Many signals are non-stationary. The power and/or spectrum of a signal can be time-varying and one example is the acoustic emission of a vessel, with variations due to changing speed and the on-off of onboard machinery. Radar signals, with frequency modulation, have a time dependent spectrum. In speech, the formant frequencies are different at different segments of a spoken word.   A complete characterization of non-stationary signals in the frequency domain must therefore include the time aspect, resulting in the time-frequency analysis of a signal.

   If the spectrum of a signal is time dependent, it is necessary to use sufficiently short segments of it (with the assumption that the spectrum is constant over each segment) to compute the spectrum.   Taking a segment of a time function is known as windowing.   As shown in Figure 1.3, this is equivalent to multiplying the signal by a window function

$$\gamma(t) = \begin{cases} 1 & , & t' \leq t \leq t' + T \\ 0 & , & otherwise \end{cases} \tag{1.50}$$

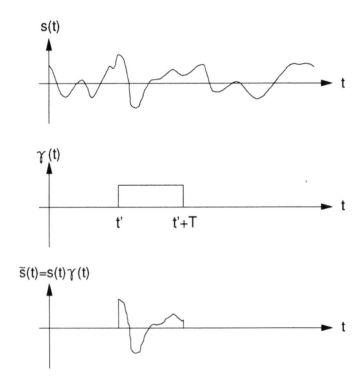

Figure 1.3 The windowing operation

The window moves along on the time axis, with overlap if desired, to produce segments of $s(t)$ for analysis. For example, we can have a 3-D plot of magnitude of the spectrum versus frequency and time, or a 2-D plot of frequency versus time, with spectral magnitudes represented by a grey scale. Such a plot is called the spectrogram in speech analysis.

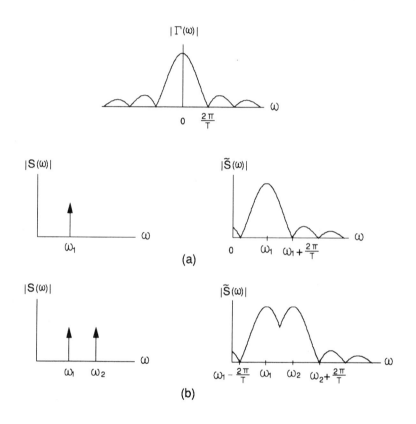

Figure 1.4 The windowed spectrum of (a) a single sinusoid, and (b) two sinusoids

A segment of length T of the signal is

$$\tilde{s}(t) = s(t)\,\gamma(t) \qquad\qquad (1.51)$$

and the Fourier transform of $\tilde{s}(t)$ is, by the convolution theorem

$$\tilde{S}(\omega) = S(\omega) * \Gamma(\omega) \qquad (1.52)$$

where $S(\omega)$ and $\Gamma(\omega)$ are the Fourier transforms of $s(t)$ and $\gamma(t)$ and $*$ denotes convolution. Suppose $s(t)$ is stationary and is an infinite duration sinusoid of frequency ω_1, then the transforms $|S(\omega)|$, $|\Gamma(\omega)|$ and $|\tilde{S}(\omega)|$ are as shown in Figure 1.4a. Because of windowing, $|\tilde{S}(\omega)|$ is $|S(\omega)|$ spread out by the window $|\Gamma(\omega)|$. Now if $s(t)$ contains two sinusoids of equal magnitude and of frequencies ω_1 and ω_2, then $|\tilde{S}(\omega)|$ is as shown in Figure 1.4b, where the spectral shape depends on the separation $|\omega_2 - \omega_1|$. If $|\omega_2 - \omega_1| \gg 2\pi/T$, $|\tilde{S}(\omega)|$ has two distinct peaks at ω_1 and ω_2. As $|\omega_2 - \omega_1|$ gets smaller, eventually only a single peak will appear. A rule of thumb for resolvability of the two sinusoids is that the observation time T should cover at least one cycle of the beat frequency, i.e.

$$|\omega_2 - \omega_1| \geq \frac{2\pi}{T} \qquad (1.53)$$

Thus the achievable frequency resolution, $\Delta f = (\omega_2 - \omega_1)/2\pi$ of a segment of length T is $1/T$. Viewing (1.53) in another way, by considering Δf as the bandwidth of the signal, then the time-bandwidth product of a signal segment must be greater than unity to provide a Δf resolution. From an information content context, large time-bandwidth product is a desirable property. To be able to differentiate the presence of two sinusoids in a signal, it is necessary to observe the signal for a sufficient length of time. If we were to estimate the frequency of a sinusoid, from the spectrum of a windowed segment, there will be a large estimation error if the segment contains only a small fraction of a cycle, especially when noise is present.

In the time-frequency analysis of a non-stationary signal, there are two conflicting requirements. The window width T must be long enough to give the desired frequency resolution but must also be short enough so as not to blur the time dependent events. If a signal contains two impulses at d sec. apart, T must be less than d sec. long to resolve the two impulses. Good resolution in time or frequency implies good localization in time or frequency. A very narrow window, ideally an impulse, gives perfect time resolution (localization) but poor frequency resolution (localization) because it has an infinite bandwidth. On the other hand, a very narrowband filter gives good localization in frequency but poor time localization because its impulse response does not decay rapidly with time. Sinusoids are perfectly local in frequency but global in time. They have non-compact support, i.e., infinite duration. Used as basis functions in Fourier analysis, they rely on cancellation to represent (synthesize) discontinuities in time. This is the cause of the Gibb's phenomenon [3]. Sinusoids are therefore not efficient in representing functions that have compact support, i.e., non-zero only for a finite duration, compared with basis functions that have compact support. Efficiency here is measured by the number of coefficients needed in the transform domain to represent a given function.

While it is possible to design window shapes [8] to optimize, or tradeoff between, time and frequency resolutions, there is a fundamental limitation on what can be done, given a fixed T. This is captured in the Uncertainty Principle [3] which states that for any transform pair $s(t)$ and $S(\omega)$,

$$\Delta_t \Delta_\omega \geq \frac{1}{2} \tag{1.54}$$

where

$$\Delta_t^2 = \frac{\int t^2 |s(t)|^2 dt}{\int |s(t)|^2 dt} \tag{1.55}$$

and

$$\Delta_\omega^2 = \frac{\int \omega^2 |S(\omega)|^2 d\omega}{\int |S(\omega)|^2 d\omega} \tag{1.56}$$

and are measures of the variations or spread of $s(t)$ and $S(\omega)$. Consider $|s(t)|^2/\int |s(t)|^2 dt$ as the probability density function of the random variable t, then (1.55) follows immediately with Δ_t^2 as the second moment of t. An interpretation of (1.54), with Δ_t and Δ_ω as the effective duration and bandwidth of signal, is: if a signal has bandwidth Δ_ω, then its duration must be longer than $1/(2\Delta_\omega)$ and vice versa. It will be seen later that the wavelet transform, through the use of different window widths, can achieve arbitrarily small (at least theoretically) Δ_t or Δ_ω, although of course not both simultaneously.

The function that meets the bound in (1.54) is the Gaussian function. Let

$$s(t) = \frac{1}{\sqrt{(2\pi)}\,\sigma_t}\, e^{-\frac{t^2}{2\sigma_t^2}} \qquad (1.57)$$

then

$$S(\omega) = e^{-\frac{\omega^2}{2\sigma_\omega^2}} \qquad (1.58)$$

i.e., the Fourier transform of $s(t)$ is also Gaussian with $\sigma_\omega^2 = 1/\sigma_t^2$ [3]. Putting $|s(t)|^2$ and $|S(\omega)|^2$ into (1.55) and (1.56) yield $\Delta_t^2 = \sigma_t^2/2$ and $\Delta_\omega^2 = \sigma_\omega^2/2$ so that

$$\Delta_t\,\Delta_\omega = \frac{\sigma_t}{\sqrt{2}}\,\frac{\sigma_\omega}{\sqrt{2}} = \frac{1}{2} \qquad (1.59)$$

and therefore $s(t)$ meets the bound.

There are several time-frequency analysis methods [9], notably the short time Fourier transform (STFT) used to produce the spectrogram in speech analysis and the Wigner-Ville distribution. All time-frequency analysis methods can be generalized by the integral

$$P(\tau,\omega) = \frac{1}{4\pi^2} \int_u \int_\lambda \int_\theta e^{-j\theta\tau - j\lambda\omega + j\theta u}\, \phi(\theta,\lambda)$$

$$s\left(u - \frac{\lambda}{2}\right) s\left(u + \frac{\lambda}{2}\right) du\, d\lambda\, d\theta \qquad (1.60)$$

where $P(\tau,\omega)$ is the intensity, at time τ and frequency ω, of the signal $s(t)$. By choosing $\phi(\theta,\lambda) = 1$ and integrating (1.60) with respect to θ, we get

$$P(\tau,\omega) = \frac{1}{2\pi} \int_u \int_\lambda s\left(u - \frac{\lambda}{2}\right) s\left(u + \frac{\lambda}{2}\right) e^{-j\lambda\omega} \delta(u - \tau) \, du \, d\lambda \qquad (1.61)$$

which is the Wigner-Ville distribution

$$W(\tau,\omega) = \frac{1}{2\pi} \int_\lambda s\left(\tau - \frac{\lambda}{2}\right) s\left(\tau + \frac{\lambda}{2}\right) e^{-j\lambda\omega} \, d\lambda \qquad (1.62)$$

Next, putting

$$\phi(\theta,\lambda) = \frac{1}{2\pi} \int_l h\left(l - \frac{\lambda}{2}\right) h\left(l + \frac{\lambda}{2}\right) e^{-j\theta l} \, dl \qquad (1.63)$$

into (1.60) and integrating with respect to θ gives

$$P(\tau,\omega) = \frac{1}{4\pi^2} \int_u \int_\lambda \int_l e^{-j\lambda\omega} s\left(u - \frac{\lambda}{2}\right) s\left(u + \frac{\lambda}{2}\right) h\left(l - \frac{\lambda}{2}\right)$$
$$h\left(l + \frac{\lambda}{2}\right) \delta(u - \tau - l) \, du \, d\lambda \, dl \qquad (1.64)$$

Further integration with respect to l gives $l = u - \tau$ and

$$P(\tau,\omega) = \frac{1}{4\pi^2} \int_u \int_\lambda e^{-j\left(u + \frac{\lambda}{2}\right)\omega} e^{j\left(u - \frac{\lambda}{2}\right)\omega} s\left(u - \frac{\lambda}{2}\right) s\left(u + \frac{\lambda}{2}\right)$$
$$h\left(u - \tau - \frac{\lambda}{2}\right) h\left(u - \tau + \frac{\lambda}{2}\right) du \, d\lambda \qquad (1.65)$$

Letting $t = u + \lambda/2$ and $t' = u - \lambda/2$ give the spectrogram, the magnitude square of the STFT:

$$|\text{STFT}(\tau,\omega)|^2 = \left| \frac{1}{2\pi} \int_t e^{-j\omega t} s(t) h(t - \tau) \, dt \right|^2 \qquad (1.66)$$

In (1.66), $h(t - \tau)$ is the window function which slides along $s(t)$ and the STFT is the Fourier transform of the product. Hence the STFT maps a single dimension signal $s(t)$ into a 2-D domain of time and frequency. When $h(t)$ is a Gaussian window, the STFT is called the Gabor transform [9]. The spectrogram plots frequency versus time, with the spectral magnitudes shown on a grey scale (Figure 1.5). It gives a display of how the spectral magnitudes of a signal change with time. For speech analysis, a typical window may be a Hamming window [10] since it requires less computations than the Gaussian window.

1.5 Applications of Time-Frequency Analysis

There are many signal processing problems, due to the time-varying nature of the signal, that require time-frequency analysis, to reveal its spectral characteristics as a function of time. The spectrogram for speech analysis is one example previously mentioned. There are wideband and narrowband spectrograms. The former has a window width typically less than $7m$ sec., which is in the order of a pitch period, and therefore has sufficient time resolution to resolve individual formants and estimate the pitch. However, its frequency resolution is insufficient to resolve the individual pitch harmonics. In the narrowband case, the window duration is about three or four times of a pitch period. Now the individual harmonics of the excitation are resolved but not the pitch periods.

In passive sonar, a spectral display of the receiver output, which may be from an array or a sonobuoy [11] provides information for detection and classification. The spectrum is both wideband, due to cavitation and hydrodynamic noise, and narrowband, for signals from the propeller and onboard machinery. This spectrum will change with speed and sea conditions and hence is time-varying. The Lofargram (low frequency analysis record gram) plots a sequence of spectra against time, giving a waterfall like display. Again there is a need for different analysis window widths. Long term record, ranging from ten to forty sec., are needed to give a good estimation of the line frequencies for classification. Short term records, below 12 sec. are used to track the spectral variations of a source. Other application examples, from geophysics to bio-medicine, are in [9]. The main point to remember is that frequency and time resolutions cannot be made small simultaneously. Different window widths are needed to meet a desired resolution in time or frequency. However, the duration over which a signal is stationary often dictates the longest width that can be used.

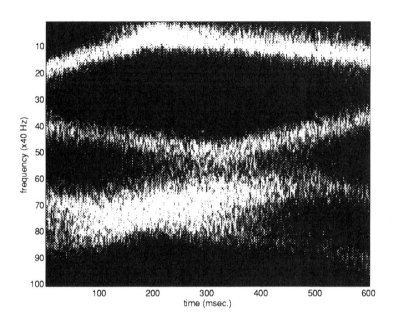

Figure 1.5 A spectrogram

Chapter 2

PRINCIPLES
OF THE WAVELET TRANSFORM

2.1 Introduction

The wavelet transform, similar to the STFT, also maps a time function into a two-dimensional function of a and τ (instead of ω and τ). The parameter a is called the scale, it scales a function by compressing or stretching it, and τ is the translation of the wavelet function along the time axis. The signal $s(t)$ is assumed to be square integrable, denoted as $s(t) \in \mathbf{L}^2(\mathbf{R})$, meaning

$$\int s^2(t)\, dt < \infty \tag{2.1}$$

Note that a dc signal is not an $\mathbf{L}^2(\mathbf{R})$ function, neither is a pure sinusoid. However, all functions of finite magnitudes and compact support are. The continuous wavelet transform of $s(t)$ is [12]

$$\mathrm{CWT}(a,\tau) = \frac{1}{\sqrt{a}} \int s(t)\, \psi\left(\frac{t-\tau}{a}\right) dt \tag{2.2}$$

where $\psi(t)$ is the basic (or mother) wavelet and $\psi((t-\tau)/a)/\sqrt{a}$ the wavelet basis functions, sometimes called baby wavelets. By a change of variable $a\,t' = t$, (2.2) becomes

$$\mathrm{CWT}(a,\tau) = \sqrt{a} \int s(a\,t')\, \psi\left(t' - \frac{\tau}{a}\right) dt' \tag{2.3}$$

showing the equivalence between scaling $\psi(t)$ in (2.2) or scaling $s(t)$ in (2.3) to obtain the wavelet transform.

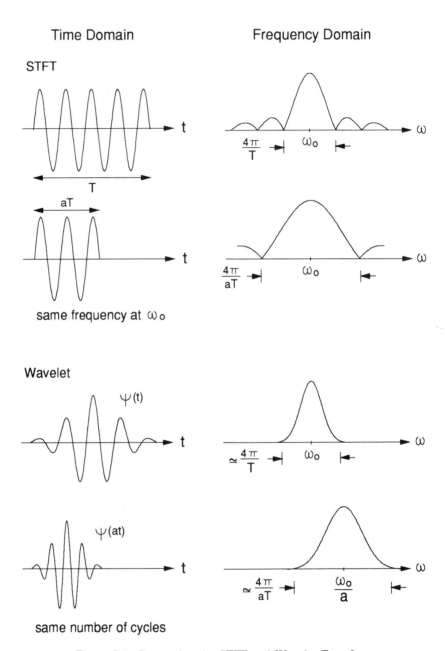

Figure 2.1 Contrasting the STFT and Wavelet Transform

There is a fundamental difference between the STFT and wavelet transform. In the STFT, at an analyzing frequency of ω_o, changing the window width in (1.66) will increase or decrease the number of cycles of ω_o inside the window. In the wavelet transform, at a carrier frequency of ω_o (see (2.3) for example), since window width changes mean dilation or compression, the carrier frequency becomes ω_o / a for a window width change from T to aT. However, the number of cycles inside the window remains constant. Figure 2.1 is an illustration of this difference. The frequency resolution is directly proportional to the window width in both the STFT and the wavelet transform. However, in the latter, a centre frequency shift necessarily accompanies a window width change (time scaling).

The basic wavelet $\psi(t)$ can be real or complex, with the resulting wavelet transform being real or complex. When $\psi(t)$ is complex, then its complex conjugate is used in (2.2) and (2.3). For some applications, it may be advantages to use complex wavelets since the phase of the wavelet transform may contain useful information. Examples of $\psi(t)$ and their Fourier transforms are:

(i)   Modulated Gaussian (Morlet)

$$\psi(t) = e^{j\omega_o t} e^{-\frac{t^2}{2}}$$

$$\Psi(\omega) = \sqrt{2\pi}\, e^{-\frac{(\omega - \omega_o)^2}{2}} \tag{2.4}$$

(ii)   Second derivative of a Gaussian

$$\psi(t) = (1 - t^2)\, e^{-\frac{t^2}{2}}$$

$$\Psi(\omega) = \sqrt{2\pi}\, \omega^2 e^{-\frac{\omega^2}{2}} \tag{2.5}$$

(iii) Haar

$$\psi(t) = \begin{cases} 1 \quad, & 0 \leq t \leq 1/2 \\ -1 \quad, & 1/2 \leq t < 1 \\ 0 \quad, & otherwise \end{cases}$$

$$\Psi(\omega) = j \, e^{-j\frac{\omega}{2}} \, \frac{\sin^2(\omega/4)}{\omega/4} \tag{2.6}$$

(iv) Shannon

$$\psi(t) = \frac{\sin(\pi t/2)}{\pi t/2} \cos\left(\frac{3\pi t}{2}\right)$$

$$\Psi(\omega) = \begin{cases} 1 & , & \pi < |\omega| < 2\pi \\ 0 & , & otherwise \end{cases} \tag{2.7}$$

They are plotted in Figure 2.2 and the scaled and translated versions of (iii) are in Figure 2.3.

From Figures 2.2 and 2.3, we can deduce the following wavelet properties:

(i) $\Psi(\omega) = 0$ at $\omega = 0$, or equivalently $\int \psi(t)\, dt = 0$, i.e., they have zero dc components.

(ii) They are bandpass signals.

(iii) They decay rapidly towards zero with time (the original French word is "Ondelette").

Property (i) is a consequence of the admissibility condition of a wavelet, the condition that ensures the wavelet transform has an inverse (see Section 2.2) and (ii) follows from (i).   The rapid decay of $\psi(t)$ is not necessary theoretically for $\psi(t)$ to be a wavelet. However, $\psi(t)$ in practice should have compact support, in order to have good time localization.

Comparing (2.2) with (1.66), we see the similarity between $\psi(t)$ of the CWT and $h(t)e^{-j\omega t}$ of the STFT. There are four ways to view the integral operation of (2.2):

(i)   It computes the inner product, or the cross-correlation of $s(t)$ with $\psi(t/a)/\sqrt{a}$, at shift τ/a. It therefore computes the 'similarity' between $s(t)$ and $\psi(t/a)/\sqrt{a}$, or the component of $s(t)$ that is 'common' to $\psi(t/a)/\sqrt{a}$.

(ii) It is the output of a bandpass filter of impulse response $\psi(-t/a)/\sqrt{a}$, of input $s(t)$, at the instant τ/a.

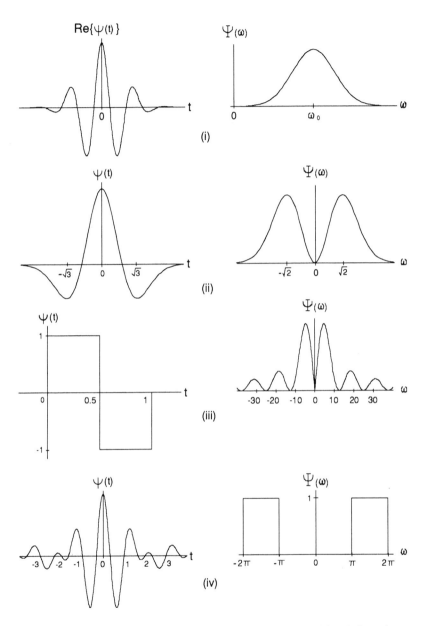

Figure 2.2 Some wavelets and their transforms. (i) modulated Gaussian (Morlet), (ii) second derivative of a Gaussian, (iii) Haar, (iv) Shannon.

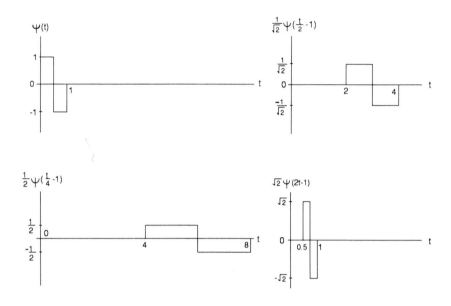

Figure 2.3   The Haar wavelet and its baby wavelets

(iii) Since (2.3) is identical to (2.2), it also computes the inner product or the cross-correlation of a scaled signal $s(at)$ with $\sqrt{a}\,\psi(t)$, at shift τ/a.

(iv) It follows from (2.3) that the CWT is also the output of a bandpass filter of impulse response $\sqrt{a}\,\psi(-t)$, of input $s(at)$, at the instant τ/a.

These different interpretations of (2.2) give rise to different implementations of the wavelet transform. The choice depends on the algorithms available and on applications. The main distinction is that in one case, the cross-correlation is between $s(t)$ and the baby wavelets and is equivalent to finding the output of a bank of bandpass filters, of impulse responses $\psi(-t/a)/\sqrt{a}$ and input $s(t)$, as shown in Figure 2.4. In the other, successively scaled versions of $s(t)$ are passed through identical bandpass filters to give the transform. Figure 2.5, which gives the same transform as Figure 2.4, appears easier to implement if there is a simple way to scale $s(t)$. By discretizing $s(t)$ and limiting the scaling to successive compressions by

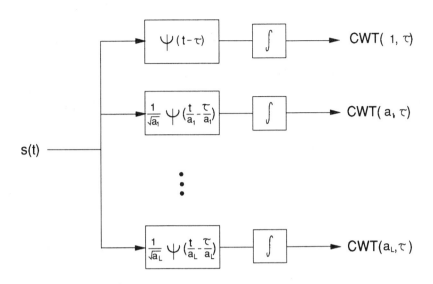

Figure 2.4 Filter bank implementation of the wavelet transform

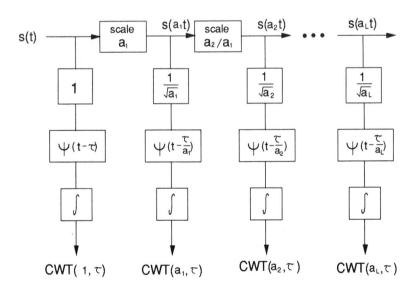

Figure 2.5 An analog of the fast wavelet transform

two, then each scale box in Figure 2.5 becomes a lowpass filter followed by decimation by two. This is the basis of the Fast Wavelet Transform, studied in Section 3.3.

There are four different types of wavelet transforms and since there are yet no standard names for them, the terminology in the literature can at times be confusing. We list them here and give names that we considered most suitable:

(i)   The continuous wavelet transform

$$\text{CWT}(a,\tau)=\frac{1}{\sqrt{a}}\int s(t)\,\psi\!\left(\frac{t-\tau}{a}\right)dt \tag{2.8}$$

has a parallel in the Fourier transform. The variable t, scale a (frequency in the Fourier transform) and shift τ are all continuous.

(ii)   The discrete parameter wavelet transform

$$\text{DPWT}(m,n)=a_o^{-\frac{m}{2}}\int s(t)\,\psi(a_o^{-m}t-n\tau_o)\,dt \tag{2.9}$$

where the parameters a, τ are discretized to $a=a_o^m$ and $\tau=n\tau_o a_o^m$ with a_o, τ_o the sampling intervals and m, n integers. Both $s(t)$ and $\psi(a_o^{-m}t)$ are still continuous. The equivalent is the Fourier series, where only frequency is the discrete parameter. For computational efficiency, $a_o=2$ and $\tau_o=1$ are commonly used, resulting in a binary dilation of 2^{-m} and a dyadic translation of $2^m n$ [13].

(iii)   The discrete time wavelet transform

$$\text{DTWT}(m,n)=a_o^{-\frac{m}{2}}\sum_k s(k)\,\psi(a_o^{-m}k-n\tau_o) \tag{2.10}$$

which is a time discretization of (2.9), with $t=kT$ and the sampling interval $T=1$. This is similar to the discrete Fourier series, where both time and frequency are discrete. Note that for $a_o=2$, there is an output only at every 2^m samples when $2^{-m}k$ is an integer.

(iv)   The discrete wavelet transform [14]

$$\text{DWT}(m,n) = 2^{-\frac{m}{2}} \sum_{k} s(k)\, \psi(2^{-m}k - n) \tag{2.11}$$

where the discrete wavelet $\psi(k)$ can be, but not necessarily, a sampled version of a continuous counterpart. That is, it is possible that $\psi(k)$ may not have a continuous time version. When $\psi(k)$ is a discretization of a $\psi(t)$, the DWT is identical to the DTWT, with the $\psi(t)$ of (2.9). In this case, a parallel for the DWT is the discrete Fourier transform.

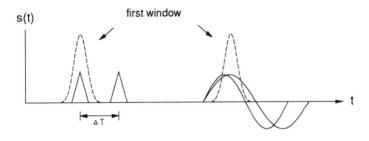

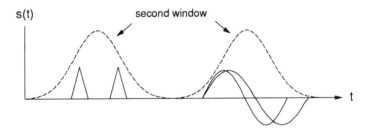

Figure 2.6 Varying time and frequency resolutions

Figure 2.6 gives a demonstration of the role of time scaling (changing window widths) in controlling time and frequency resolution in signal analysis. Let $s(t)$ contain two triangular peaks at a time separation of ΔT and two sinusoids at a frequency separation of Δf, with $\Delta f < \Delta T$. The first window has width less than ΔT and can therefore resolve the two peaks, i.e., indicate where in time each peak occurs, but cannot resolve (see (1.53)) the two sinusoids. On the other hand, the second window, which has a width greater than ΔT, can resolve the sinusoids but not the peaks. By choosing different scales for the baby wavelets in (2.2), the wavelet transform can achieve any desired resolution in time or frequency. There is no violation of the uncertainty principle: $\Delta_t \Delta_\omega$ is still lower bounded by $1/2$ as in (1.54). By time scaling, Δ_t can be arbitrarily small, with Δ_ω correspondingly large to maintain (1.54) and vice versa. Let $\psi(t)$ have $\Delta_t = \delta_t$ and $\Delta_\omega = \delta_\omega$ as in (1.55) and (1.56). Then $\psi(t/a)$ will have $\Delta_t = a\,\delta_t$ and $\Delta_\omega = \delta_w/a$. By varying a, various Δ_t and Δ_ω can be obtained but their product remains constant.

2.2 The Continuous Wavelet Transform

In any transform application, it is important that the transform have an inverse. Perfect reconstruction is a basic requirement in image coding, even though after compression, the reconstructed image will only be an approximation of the original. Template matching or pattern recognition systems often use transforms for feature extraction. These transforms should be reversible to ensure a unique representation of a pattern by a transform. If a transform is not reversible, it is conceivable to have dissimilar patterns possessing identical transforms. As a trivial example, consider the transformation $y = A\,x$, where the transformation matrix

$$A = \begin{bmatrix} 2 & 3 \\ 1 & 1.5 \end{bmatrix}$$

does not have an inverse. Then both the vectors $x = [1 \quad 2]^T$ or $[4 \quad 0]^T$ will have identical transforms of $y = [8 \quad 4]^T$.

Let $s(t) \in \mathbf{L}^2(\mathbf{R})$ and the CWT of $s(t)$, with respect to a wavelet $\psi(t)$, be

$$\mathrm{CWT}(a,\tau) = \frac{1}{\sqrt{a}} \int s(t)\, \psi\left(\frac{t-\tau}{a}\right) dt \tag{2.12}$$

If the $\psi(t)$ is such that this transform is invertible, then

$$s(t) = \frac{1}{c_\psi} \int_{-\infty}^{\infty} \int_{a>0}^{\infty} \text{CWT}(a,\tau) \frac{1}{\sqrt{a}} \psi\left(\frac{t-\tau}{a}\right) \frac{1}{a^2} da \, d\tau \qquad (2.13)$$

where c_ψ is a constant that depends only on $\psi(t)$ and a is positive. The constant has value

$$c_\psi = \int_0^\infty \frac{|\Psi(\omega)|^2}{\omega} d\omega < \infty \qquad (2.14)$$

which in turn imposes an admissibility condition on $\psi(t)$. For $c_\psi < \infty$, $\psi(t)$ must be such that

$$|\Psi(\omega)| < \infty \quad , \qquad \text{for any } \omega \qquad (2.15)$$

and $\Psi(0) = 0$, implying that

$$\int \psi(t) \, dt = 0 \qquad (2.16)$$

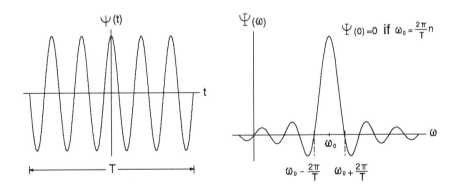

Figure 2.7   A truncated sinusoid as a wavelet

meaning that $\psi(t)$ cannot have non-zero dc. Returning to Figure 2.2, which plots several $\Psi(\omega)$, it is seen that they all satisfy the admissibility condition. For the modulated Gaussian, $\Psi(0)$ is not exactly zero, although by choosing

ω_o sufficiently large, ($\omega_o > 5.5$ say), $\Psi(0) \approx 0$. A truncated segment of a sinusoid will also qualify as a wavelet if its frequency $\omega_o = 2n\pi/T$, $n = 1, 2, \dots$ or if ω_o is sufficiently large (see Figure 2.7). The proof of the invertability of the CWT, together with condition (2.14), are obtained by applying the resolution of identity theorem, given in Appendix A. This theorem states that the transformation of a one-dimensional signal $s(t)$ into the two-dimensional wavelet domain of (a, τ) via (2.12) is invertible if the transformation is an isometry (energy preserving) up to a constant factor c_ψ, given by (2.14). A special case of the resolution of identity is Parseval's theorem with $c_\psi = 2\pi$.

Figure 2.8 and 2.9 give two plots of the CWT of an impulse and a signal of sinusoids. In the first case, $s(t) = \delta(t - 4)$, an impulse occuring at $4 \sec$. In the other,

$$s(t) = \begin{cases} \sin(5.89t) & , \quad 0 \le t \le 1 \\ \sin(8.83t) & , \quad 1 \le t \le 2 \\ \sin(5.89t) + \sin(8.83t) & , \quad 2 \le t \le 3 \\ 0 & , \quad t \ge 3 \end{cases} \qquad (2.17)$$

The wavelet is

$$\psi(t) = e^{-j5.3t} \, e^{-\frac{t^2}{2}} \qquad (2.18a)$$

with the baby wavelets

$$\frac{1}{\sqrt{a}} \psi\left(\frac{t-\tau}{a}\right) = \frac{1}{\sqrt{a}} e^{-j5.3\left(\frac{t-\tau}{a}\right)} e^{-\frac{1}{2}\left(\frac{t-\tau}{a}\right)^2} \qquad (2.18b)$$

Since $\psi(t)$ is complex, both magnitude squared $|CWT(a,\tau)|^2$, called the scalogram [15] and phase of the CWT are available. Figure 2.8(a), which plots magnitude of the $CWT(a,\tau)$ against a and τ, shows an impulse at $\tau = 4$ and $a \approx 0$. When $a \approx 0$, the corresponding baby wavelet approximates an impulse and hence has a high correlation with the signal. The phase plot in Figure 2.8(b) has lines of constant phase converge towards the impulse location at $a \approx 0$ and $\tau = 4$. This particular property is useful for detecting discontinuity in a signal and section 4.3 has more discussion on this subject.

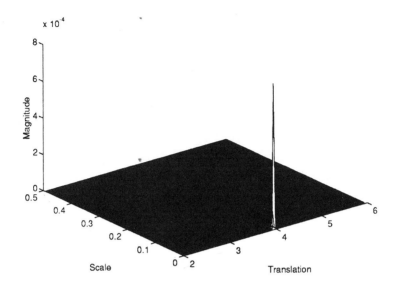

Figure 2.8(a) Magnitude of the CWT of an impulse

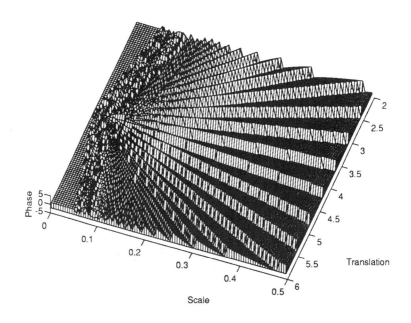

Figure 2.8*(b)*   Phase of the CWT of an impulse

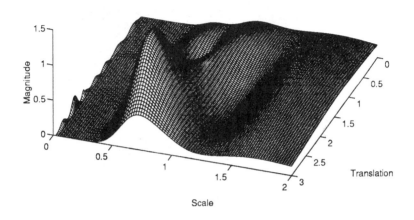

Figure 2.9(a) Magnitude of the CWT of sinusoids

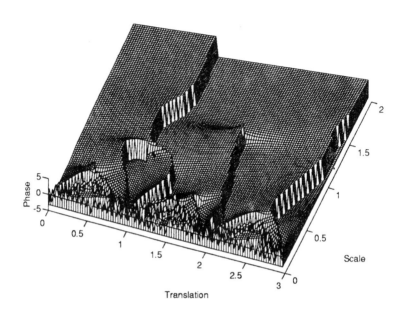

Figure 2.9*(b)*   Phase of the CWT of sinusoids

For the second signal, a sinusoid of $\omega = 5.89$ has maximum correlation with the baby wavelet when $a = 5.3/5.89 = 0.9$ and similarly $a = 0.6$ for $\omega = 8.83$. These two values of a are where peaks appear in Figure 2.9(a), indicating how the time scale a relates to frequency. This diagram, however, does not give a clear time localization of the sinusoids. Their transition times are better resolved instead in the phase plot of Figure 2.9(b). Note that if the CWT(a, τ) uses a different mother wavelet, the plots could have been quite different.

2.3 Discrete Parameter Wavelet Transform

In (2.8), both (a, τ) are continuous variables and there is a redundancy in the CWT representation of $s(t)$. There is certainly no need to compute CWT(a, τ) for all possible (a, τ). Additionally, it is of practical necessity that (a, τ) take on only a finite number of values. With (a, τ) discrete, is a complete representation of $s(t)$ still possible? That is, can $s(t)$ be recovered from CWT(a, τ) with (a, τ) discrete? The answer is yes, provided that the sampling of (a, τ) is sufficiently dense. This should come as no surprise since there is already a parallel in the perfect recovery of a signal from its samples taken at or above the Nyquist rate. Depending on the type of $\psi(t)$ and the sampling grid of (a, τ), sometimes duals (Section 1.3) are required for perfect reconstruction. A special case occurs when (a, τ) are samples of a dyadic grid (see discussion following (2.9)), when certain $\psi(t)$ can produce orthonormal $\psi((t-\tau)/a)/\sqrt{a}$, with (a, τ) discrete (see below and Chapter 3). Consequently, $s(t)$ can be exactly synthesized as a weighted sum of these orthonormal basis functions.

When (a, τ) is discrete and given by

$$a = a_o^m \quad , \quad \tau = n\tau_o a_o^m \quad , \quad m, n \text{ integer} \tag{2.19}$$

the discrete parameter wavelet transform is [4]

$$\text{DPWT}(m, n) = \int s(t)\, \psi_{mn}(t)\, dt \tag{2.20}$$

where

$$\psi_{mn}(t) = a_o^{-\frac{m}{2}} \psi(a_o^{-m} t - n\tau_o) \quad , \quad \psi_{00}(t) = \psi(t) \tag{2.21}$$

and a_o and τ_o are constants that determine the sampling intervals. Note that both $s(t)$ and $\psi(t)$ are still continuous functions of time. From the discussion in the previous paragraph, it is of interest to study the type of $\psi(t)$ and sampling intervals for (a,τ) that permit perfect reconstruction

$$s(t) = c \sum_m \sum_n \text{DPWT}(m,n) \, \psi_{mn}(t) \qquad (2.22)$$

where c is some constant dependent only on $\psi(t)$.

Suppose there is oversampling, then (2.22) will not only hold but moreover, it is possible to have a non-unique representation of $s(t)$, with respect to the same $\psi(t)$. Let $\text{DP}\widetilde{\text{W}}\text{T}(m,n)$ be the DPWT due to $(\bar{a}_o, \tilde{\tau}_o)$ and $\text{DP}\widehat{\text{W}}\text{T}(m,n)$ due to $(\hat{a}_o, \hat{\tau}_o)$. Since there is redundancy, even though $\text{DP}\widetilde{\text{W}}\text{T}(m,n) \neq \text{DP}\widehat{\text{W}}\text{T}(m,n)$, we can still have

$$s(t) = \bar{c} \sum_m \sum_n \text{DP}\widetilde{\text{W}}\text{T}(m,n) \, \tilde{\psi}_{mn}(t)$$

$$= \hat{c} \sum_m \sum_n \text{DP}\widehat{\text{W}}\text{T}(m,n) \, \hat{\psi}_{mn}(t) \qquad (2.23)$$

where $\tilde{\psi}_{mn}(t)$ and $\hat{\psi}_{mn}(t)$ are as defined in (2.21) corresponding to $(\bar{a}_o, \tilde{\tau}_o)$ and $(\hat{a}_o, \hat{\tau}_o)$. Redundancy from oversampling permits two distinct sets of baby wavelets, produced from the same mother wavelet, to give exact but distinct synthesis of $s(t)$. In contrast, if the sampling is sparse, (2.22) will not hold.

Daubechies [4,12] applied the theory of frames to put down formally the reconstruction conditions for the DPWT. Section 1.3 contains a discussion on the principles of frames in the context of vectors. Without proof, they are now extended to time functions $\psi_{mn}(t)$. The goal is to develop conditions for $\psi_{mn}(t)$ that permit perfect reconstruction, via duals if necessary. For convenience, let $<\psi_{mn}(t), \psi_{mn}(t)> = 1$ and let

$$\text{DPWT}(m,n) = <s(t), \psi_{mn}(t)> = \int s(t) \, \psi_{mn}(t) \, dt \qquad (2.24)$$

Suppose the $\psi_{mn}(t)$ do not constitute an orthonormal set so that (2.22) is not valid. Then what are the conditions on the $\psi_{mn}(t)$ that would allow the alternate reconstruction

$$s(t) = \sum_m \sum_n DPWT(m,n) \, \tilde{\psi}_{mn}(t) \qquad (2.25)$$

by duals $\tilde{\psi}_{mn}(t)$? By mimicking the development of Section 1.3, with time functions replacing vectors, it is easy to see that for (2.25) to hold, the $\psi_{mn}(t)$ must be elements of a frame. The $\tilde{\psi}_{mn}(t)$ are then elements of the dual frame. If the $\psi_{mn}(t)$ form a frame, they must obey the inequality

$$A \, \| s(t) \|^2 \le \sum_m \sum_n | <s(t), \psi_{mn}(t)> |^2 \le B \, \| s(t) \|^2 \qquad (2.26)$$

with $0 < A \le B < \infty$. The constants A, B are frame bounds and are dependent only on the $\psi_{mn}(t)$, and

$$\| s(t) \|^2 = \int | s(t) |^2 \, dt < \infty \qquad (2.27)$$

because $s(t) \in \mathbf{L}^2(\mathbf{R})$.

The bounds in (2.26) ensure that the   reconstruction is numerically stable, in the sense that if $<s_1(t), \psi_{mn}(t)>$ and $<s_2(t), \psi_{mn}(t)>$ are 'close', then $s_1(t)$ and $s_2(t)$ should be 'close' as well. This follows since

$$A \, \| s_1(t) - s_2(t) \|^2 \le \sum_m \sum_n | <s_1(t) - s_2(t), \psi_{mn}(t)> |^2$$

$$\le B \, \| s_1(t) - s_2(t) \|^2 \qquad (2.28)$$

so that if $s_1(t) \approx s_2(t)$, $<s_1(t), \psi_{mn}(t)>$ must also be approximately equal to $<s_2(t), \psi_{mn}(t)>$. Daubechies [4] has shown that the STFT with a Gaussian window, i.e. the Gabor transform, is numerically unstable because its basis functions do not constitute a frame. In general, the discrete time STFT (both signal and the frequency variable are discrete) has an inverse if certain conditions are met [8]. While it is relatively easy to find the dual frame in the case of vectors (see (1.36)), it is not the same for functions. The formula [12] for duals, when $A \approx B$, is

$$\tilde{\psi}_{mn}(t) = \frac{2}{A+B} \psi_{mn}(t)$$

$$+ \frac{2}{A+B} \left(\psi_{mn}(t) - \frac{2}{A+B} \sum_j \sum_k \psi_{jk}(t) < \psi_{jk}(t) , \psi_{mn}(t) > \right)$$

$$+ O\left(\left(\frac{B}{A} - 1 \right)^2 \right) \tag{2.29}$$

or

$$\tilde{\psi}_{mn}(t) = \frac{2}{A+B} \psi_{mn}(t) + O\left(\frac{B}{A} - 1 \right) \tag{2.30}$$

The term $O((B/A-1)^2)$ means that it approaches zero as fast as $(B/A-1)^2$. It is not easy either to compute A or B given $\psi_{mn}(t)$ although good estimates are available in special cases [4,12].

The following is a summary of the results of applying frame theory to the study of the DPWT, in particular to its invertability:

(i)   Choice of (a_o, τ_o) in sampling the continuous variables $a = a_o^m$, $\tau = n \tau_o a_o^m$ determines the invertability of a DPWT.   Too coarse a sampling grid obviously will not allow perfect reconstruction.

(ii) There are threshold values (a_o, τ_o), for a given $\psi(t)$, below which the $\psi_{mn}(t)$ will always form a frame.   Then reconstruction is possible via (2.25).   However, computation of the $\tilde{\psi}_{mn}(t)$ in general requires recursion [4,12], unless the frame bounds $A \approx B$.

(iii) It is possible to select (a_o, τ_o) so that $A \approx B$.   Then from (2.30)

$$s(t) \approx \frac{2}{A+B} \sum_m \sum_n < s(t) , \psi_{mn}(t) > \psi_{mn}(t) \tag{2.31}$$

The closer is A to B, the better is this approximation [12].   Computation of A and B are difficult in general.   Estimates of A, B are available [4] for certain special $\psi(t)$.   Indeed there is no general procedure on the selection of (a_o, τ_o) that leads to $A \approx B$ for any $\psi(t)$.   A frame is tight if $A = B$, then

$$s(t) = \frac{1}{A} \sum_m \sum_n <s(t), \psi_{mn}(t)> \psi_{mn}(t) \qquad (2.32)$$

and if $A = B = 1$, then the $\psi_{mn}(t)$ are orthonormal basis functions (see Chapter 3). This is easy to verify by letting the $\psi_{mn}(t)$ in (2.29) be orthonormal to show that $\tilde{\psi}_{mn}(t) = \psi_{mn}(t)$.

A practical sampling scheme is $a = 2^m$, $\tau = n \, 2^m$, i.e., $a_o = 2$ and $\tau_o = 1$ in (2.19) so that

$$\psi_{mn}(t) = 2^{-\frac{m}{2}} \psi(2^{-m} t - n) \qquad (2.33)$$

with this octave time scaling and dyadic translation, the sampled values of (a, τ) are as shown in the dyadic grid of Figure 2.10. Since the Fourier transform of $\psi(at)/\sqrt{a}$ is $\Psi(\omega/a)/a\sqrt{a}$, the centre frequency and bandwidth of a wavelet are both scaled by $1/a$ for a time scaling of a. Thus the Q of all baby wavelets

$$Q = \frac{\text{centre frequency}}{\text{bandwidth}} = constant \qquad (2.34)$$

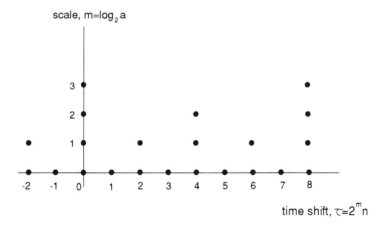

Figure 2.10   The dyadic sampling grid

giving rise to the so called constant-Q analysis capability of wavelets. With a changing by octaves, the centre frequency and bandwidth also change by octaves. The frequency resolution decreases (increasing bandwidth) with increasing centre frequencies. Returning to Figure 2.4, which is a filter bank implementation of the wavelet transform, it is seen that octave scaling provides a regular spread of the filter bank centre frequencies in a logarithmic scale (as opposed to the linear scale for the STFT). It is reasonable to expect a measurement system to possess a resolution that is a constant percentage of the quantity being measured. A constant-Q frequency discriminator can resolve $10000\,Hz$ from $11000\,Hz$ ($1000\,Hz$ resolution) and $100\,Hz$ from $110\,Hz$ ($10\,Hz$ resolution). The resolution as a percentage of the centre frequency is identical in both instances of 10%. However, the latter has a finer resolution of $10\,Hz$ compared with $1000\,Hz$ for the former. Some physiological systems, for example, the cochlea [16] and the retina [1], exhibit such properties. The narrowband processing of sonar signals sometimes also employs a constant-Q analysis. Octave processing [17] gives a vernier spectral analysis of a signal. For detection centered at f_c with a resolution (bandwidth) of Δf, the signal is sampled at f_s and mixed to baseband. Then it is sampled at $f_s/2$ for detection at $f_c/2$ with a resolution of $\Delta f/2$, etc., as shown in Figure 2.11. This analysis sampling time by two is easily obtained by filtering and then decimation by two. Since a digital filter of bandwidth Δf at a sampling frequency of f_s is identical to one of $\Delta f/2$ at $f_s/2$, all the filters are identical.

When the $\psi_{mn}(t)$ are orthonormal, then

$$< \psi_{mn}(t) , \psi_{kl}(t) > = \delta_{m-k}\, \delta_{n-l} \tag{2.35}$$

The search for discrete parameters orthonormal wavelets with compact support has been a subject of intense research [13,18] due to their obvious desirable properties and potential for wide applications. As will be seen in Chapter 3, the only non-trivial ones are those of Daubechies [18], constructed by recursion. There is no analytic expression for them. Two examples of orthonormal, discrete parameters (dyadic sampling) follow. The first one has compact support in time while the second does not.

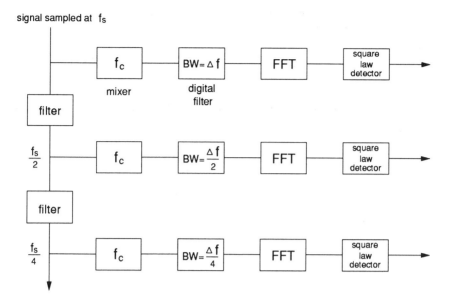

Figure 2.11   Octave processing of signals

(i)   The Haar Wavelets
      The Haar wavelet

$$\psi(t) = \begin{cases} 1 & , \quad 0 \le t < 0.5 \\ -1 & , \quad 0.5 \le t \le 1 \\ 0 & , \quad otherwise \end{cases} \tag{2.36}$$

is a special case of the Daubechies wavelets [18]. A plot of $\psi(t)$ and $\Psi(\omega)$ is in Figure 2.2. Now

$$\psi_{mn}(t) = 2^{-\frac{m}{2}} \psi(2^{-m} t - n)$$

$$= \begin{cases} 1 & , \quad 2^m n \le t < 2^m n + 2^{m-1} \\ -1 & , \quad 2^m n + 2^{m-1} \le t < 2^m n + 2^m \\ 0 & , \quad otherwise \end{cases} \tag{2.37}$$

and Figure 2.3 contains samples of these wavelets for various m and n. It is easy to verify graphically that

$$\int \psi_{mn}(t)\,\psi_{kl}(t)\,dt = \begin{cases} 0 & , \quad m \neq k \quad or \quad n \neq l \\ 1 & , \quad m = k \quad and \quad n = l \end{cases} \tag{2.38}$$

Since these $\psi_{mn}(t)$ are orthonormal, they must satisfy (2.32) and (2.26) with $A = B = 1$. To show this, let

$$s(t) = \begin{cases} a & , \quad 0 \leq t < 0.5 \\ b & , \quad 0.5 \leq t < 1 \\ 0 & , \quad otherwise \end{cases} \tag{2.39}$$

For this $s(t)$,

$$<s(t), \psi_{mn}(t)> = 0 \quad , \quad m < 0 \quad or \quad n \neq 0 \tag{2.40}$$

and

$$<s(t), \psi_{00}(t)> = \frac{a}{2} - \frac{b}{2}$$

$$<s(t), \psi_{10}(t)> = \frac{1}{\sqrt{2}}\left(\frac{a}{2} + \frac{b}{2}\right)$$

$$<s(t), \psi_{20}(t)> = \frac{1}{2}\left(\frac{a}{2} + \frac{b}{2}\right)$$

$$\vdots$$

$$<s(t), \psi_{m0}(t)> = \frac{1}{2^{\frac{m}{2}}}\left(\frac{a}{2} + \frac{b}{2}\right) \tag{2.41}$$

so that

$$<s(t), \psi_{00}(t)> \psi_{00}(t) = \begin{cases} \dfrac{a}{2} - \dfrac{b}{2} & , \quad 0 \leq t < 0.5 \\[2mm] -\dfrac{a}{2} + \dfrac{b}{2} & , \quad 0.5 \leq t < 1 \end{cases} \tag{2.42}$$

and for $m \geq 1$

$$<s(t), \psi_{m0}(t)> \psi_{m0}(t) = \frac{1}{2^m}\left(\frac{a}{2}+\frac{b}{2}\right) \quad , \qquad 0 \leq t \leq 1 \qquad (2.43)$$

Hence

$$\sum_{m=0}^{\infty} <s(t), \psi_{m0}(t)> \psi_{m0}(t)$$

$$= \begin{cases} \dfrac{a}{2}-\dfrac{b}{2}+\left(\dfrac{a}{2}+\dfrac{b}{2}\right)\displaystyle\sum_{m=1}^{\infty}\dfrac{1}{2^m} \quad , & 0 \leq t < 0.5 \\[4mm] -\dfrac{a}{2}+\dfrac{b}{2}+\left(\dfrac{a}{2}+\dfrac{b}{2}\right)\displaystyle\sum_{m=1}^{\infty}\dfrac{1}{2^m} \quad , & 0.5 \leq t < 1 \end{cases} \qquad (2.44)$$

But

$$\sum_{m=1}^{\infty} \frac{1}{2^m} = 1 \qquad (2.45)$$

giving

$$\sum_{m=0}^{\infty} <s(t), \psi_{m0}(t)> \psi_{m0}(t) = \begin{cases} a \quad , & 0 \leq t < 0.5 \\ b \quad , & 0.5 \leq t < 1 \end{cases} \qquad (2.46)$$

which verifies (2.32). On using (2.45) again, the sum

$$\sum_{m=0}^{\infty} |<s(t), \psi_{m0}(t)>|^2 = \frac{a^2}{2}+\frac{b^2}{2} = \int_0^1 s^2(t)\, dt \qquad (2.47)$$

confirming (2.26). As any $s(t)$ can be modelled as a succession of these two piece-wise constant segments, with arbitrarily small widths if necessary (a width of 2^m can be made arbitrarily small with m sufficiently negative), the above verification can be generalized to hold for any $s(t)$.

(ii) The Shannon Wavelets
The mother wavelet is

$$\psi(t) = \frac{\sin(\pi t/2)}{\pi t/2} \cos\left(\frac{3\pi t}{2}\right) \tag{2.48}$$

with a spectrum as shown in Figure 2.2. The baby wavelets, with dyadic sampling on the parameters (a, τ), are

$$\psi_{mn}(t) = 2^{-\frac{m}{2}} \frac{\sin(\pi(2^{-m}t - n)/2)}{\pi(2^{-m}t - n)/2} \cos\left(\frac{3\pi}{2}(2^{-m}t - n)\right) \tag{2.49}$$

with the corresponding spectra

$$\Psi_{mn}(\omega) = \begin{cases} 2^{m/2} e^{-j\omega n 2^m} &, \quad 2^{-m}\pi < |\omega| < 2^{-m} 2\pi \\ 0 &, \quad otherwise \end{cases} \tag{2.50}$$

Clearly $\Psi_{mn}(\omega)$ and $\Psi_{kl}(\omega)$ do not overlap for $k \neq m$. Invoking the inner product preserving property [3] of the Fourier transform, which states that

$$<x(t), y(t)> = \int x(t) y^*(t) dt$$

$$= \frac{1}{2\pi} \int X(\omega) Y^*(\omega) d\omega$$

$$= <X(\omega), Y(\omega)> \tag{2.51}$$

it follows that

$$<\psi_{mn}(y), \psi_{kn}(t)> = 0 \quad, \qquad k \neq m \tag{2.52}$$

Further

$$<\psi_{mn}(t), \psi_{ml}(t)> = \frac{2^m}{2\pi} \int_{2^{-m}\pi}^{2^{-m+1}\pi} 2\cos(\omega(n-l)2^m) d\omega = \delta_{n-l} \tag{2.53}$$

Hence the $\psi_{mn}(t)$ are orthonormal wavelets.

The above two wavelets are opposite of each other in terms of their localizing properties. The Haar wavelet has good time localization but poor frequency localization. Its spectrum is non-zero for $\omega \to \infty$. It does not have compact support in the frequency domain. In contrast, the Shannon wavelet has non-compact support in time and decays only as fast as $1/t$, hence it has poor time localization. Its frequency localization is good because it has the spectrum of an ideal bandpass filter. There are orthonormal wavelets that are between these two types, giving both acceptable localizations in time and frequency, as will be shown in the next Chapter.

Appendix A - Resolution of the Identity

Theorem: Let

$$\text{CWT}(a,\tau) = \frac{1}{\sqrt{a}} \int s(t)\, \psi\!\left(\frac{t-\tau}{a}\right) dt \qquad (A1)$$

If $s(t) \in \mathbf{L}^2(\mathbf{R})$ and $\psi(t)$ is such that

$$c_\psi = \int \frac{|\Psi(\omega)|^2}{\omega}\, d\omega < \infty \qquad (A2)$$

then

$$s(t) = \frac{r(t)}{c_\psi} \qquad (A3)$$

with

$$r(t) = \frac{1}{\sqrt{a}} \int \int \text{CWT}(a,\tau)\, \psi\!\left(\frac{t-\tau}{a}\right) \frac{1}{a^2}\, da\, d\tau \qquad (A4)$$

Proof:

By definition,

$$S(\omega) = \int s(t)\, e^{-j\omega t}\, dt \qquad (A5)$$

and

$$\Psi(\omega) = \int \psi(t)\, e^{-j\omega t}\, dt \qquad (A6)$$

From (A6), with $t = a\,t'$

$$\int \psi\!\left(\frac{t-\tau}{a}\right) e^{-j\omega t}\, dt = \int \psi\!\left(t' - \frac{\tau}{a}\right) e^{-ja\omega\left(t' - \frac{\tau}{a}\right)} e^{-j\omega\tau}\, d(a t') \qquad (A7)$$

$$= a\, \Psi(a\omega)\, e^{-j\omega\tau} \qquad (A8)$$

or

$$\psi\left(\frac{t-\tau}{a}\right)=\frac{a}{2\pi}\int \Psi(a\omega)\,e^{-j\omega\tau}e^{j\omega t}\,d\omega \qquad (A9)$$

In $(A1)$, replacing $s(t)$ by its inverse Fourier transform gives

$$\text{CWT}(a,\tau)=\frac{1}{\sqrt{a}}\int\frac{1}{2\pi}\int S(\omega)\,e^{j\omega t}\,d\omega\,\psi\left(\frac{t-\tau}{a}\right)dt \qquad (A10)$$

or

$$\text{CWT}(a,\tau)=\frac{1}{2\pi\sqrt{a}}\int S(\omega)\int\psi\left(\frac{t-\tau}{a}\right)e^{ja\omega(t-\tau)/a}\,e^{j\omega\tau}\,dt\,d\omega \qquad (A11)$$

which, on using $(A7)$ becomes

$$\text{CWT}(a,\tau)=\frac{\sqrt{a}}{2\pi}\int S(\omega)\,\Psi^{*}(a\omega)\,e^{-j\omega\tau}\,d\omega \qquad (A12)$$

Substituting $(A12)$ and $(A9)$ into $(A4)$ gives

$$r(t)=\frac{1}{4\pi^{2}}\int\int\int a\,S(\omega)\,\Psi^{*}(a\omega)\,e^{j\omega\tau}\,d\omega$$

$$\int\Psi(a\Omega)\,e^{-j\Omega\tau}e^{j\Omega t}\,d\Omega\frac{1}{a^{2}}da\,d\tau \qquad (A13)$$

But

$$\frac{1}{2\pi}\int e^{j(\omega-\Omega)\tau}\,d\tau=\delta(\omega-\Omega) \qquad (A14)$$

hence

$$r(t)=\frac{1}{2\pi}\int\int\int\frac{1}{a}S(\omega)\,\Psi^{*}(a\omega)\,\Psi(a\Omega)\,\delta(\omega-\Omega)\,e^{j\Omega t}\,d\omega\,d\Omega\,da \qquad (A15)$$

$$=\frac{1}{2\pi}\int S(\omega)\,e^{j\omega t}\,d\omega\int\Psi^{*}(a\omega)\,\Psi(a\omega)\frac{1}{a\omega}d(a\omega) \qquad (A16)$$

Letting $\tilde{\omega} = a\omega$ in the second term of $(A16)$ gives

$$r(t) = c_\psi \, s(t) \qquad (A17)$$

where

$$c_\psi = \int \frac{|\Psi(\tilde{\omega})|^2}{\tilde{\omega}} \, d\tilde{\omega} \qquad (A18)$$

Chapter 3

MULTIRESOLUTION ANALYSIS, WAVELETS AND DIGITAL FILTERS

3.1 Introduction

The idea of multiresolution analysis (MRA) is very similar to subband decomposition and coding, where for coding efficiency, a signal is divided into a set of frequency bands [2]. In Figure 3.1, the sequence $s(n)$, $n = 0, 1, \ldots, N-1$, is bandlimited from 0 to 1. With lowpass ($\overline{g}(l)$) and highpass ($\overline{h}(l)$) filters and decimation, the scheme in Figure 3.1 decomposes $s(n)$ into its subband components of $d_n^1, d_n^2, \ldots, d_n^L$ and c_n^L, representing the coarsest, conceivably the dc, as $L \to \infty$, component. The symbols $\downarrow 2$ and $\uparrow 2$ stand for subsampling and upsampling by two, respectively. Subsampling by two is dropping every other point in the sequence and renumbering the sequence. Upsampling by two is inserting zeros between points in a sequence and then renumbering. In Figure 3.1, there is no aliasing from subsampling by two the output of each filter, since each filter reduces the bandwidth of its input by two. In speech or image coding, a different coder is assigned to each subband component. For example, there are typically 16 bands in image coding [1] and the number of bits per pixel is higher for coding the lower frequency components which have less points than the higher components due to subsampling.

Assuming no coders for simplicity, a receiver can reconstruct the original $s(n)$ from the subband components as shown in Figure 3.2. Upsampling is followed by a bandpass or lowpass filter, to eliminate aliasing. In Figure 3.1, the bands are in contiguous octaves. Any other band divisions, possibly with overlap, will also work. This octave division corresponds to the Laplacian pyramid [2] as well as the MRA scheme of Mallat [19]. Besides giving a constant-Q analysis, it is also critically sampled, i.e., the sum of data rates at all output d_n^m, $m = 1, 2, \ldots, L$ and c_n^L, equals the input data rate.

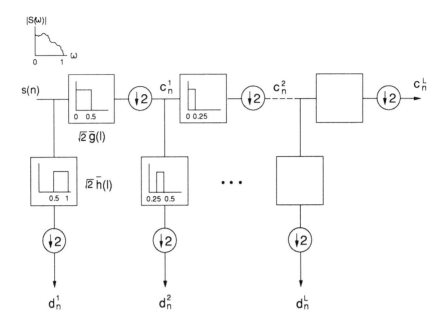

Figure 3.1 Multiresolution analysis

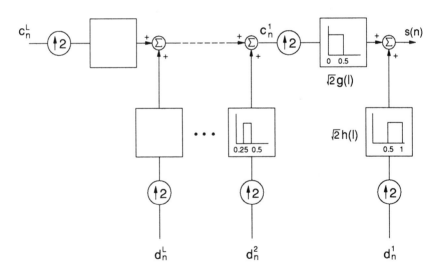

Figure 3.2 Reconstruction from subband components

The resolution of a signal is a qualitative term related to its frequency content. For a lowpass signal, the lower its frequency content (the narrower the bandwidth), the smaller is its resolution. Viewing subband decomposition as an MRA, the frequency content or resolution of the sequence c_n^m, output of lowpass filters, decreases with increasing m, until it reaches the dc component c_n^∞, of lowest resolution. The highpass (or bandpass) output d_n^m are called the high frequency detail, or difference, which is the difference in resolution between c_n^{m-1} and c_n^m.

Downsampling by two, an operation which drops every other point of a sequence and then renumbers the decimated sequence, is equivalent to compressing the sequence by two. However, since lowpass filtering must precede downsampling, it also halves the signal bandwidth and reduces its resolution. Since the DPWT with dyadic sampling on the (a, τ) parameters also scales a sequence successively by two, one might expect a link between the DPWT and the MRA of a signal. Indeed, it will be seen in the sequel that the d_n^m are the DPWT(m, n) of $s(t)$, with respect to an orthonormal, compactly supported set of $\psi_{mn}(t)$. However, the order and coefficients of the lowpass and bandpass filters determine the $\psi_{mn}(t)$, not the reverse. As yet, there is no theory available to design the filters to implement a chosen set of $\psi_{mn}(t)$.

3.2 The DPWT and MRA

To establish a connection between MRA and the DPWT, consider Figure 3.3, which is the first stage of the MRA decomposition of Figure 3.1. Let $s(n)$ be the samples of $f(\lambda)$, which is the inner product of $s(t)$ and a function $\phi(t - \lambda)$ so that, for $\lambda = nT, T = 1$,

$$s(n) = f(\lambda = nT) = \int s(t) \phi(t - n) \, dt \qquad (3.1)$$

Thus $s(n)$ are the samples of the inner product of $s(t)$ and $\phi(t - \lambda)$. They are not the samples of $s(t)$. The inner product operation puts the $s(n)$ in the proper subspace, spanned by $\phi(t - n)$, so that the details d_n^m are indeed the DPWT(m, n) of $s(t)$. The necessity of this operation will become clear and $\phi(t)$ will be defined, in the derivations to follow.

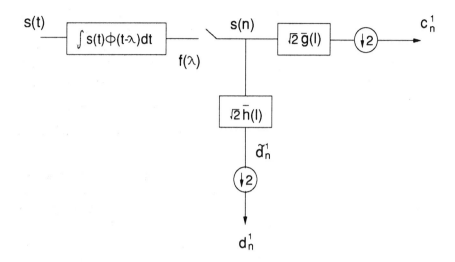

Figure 3.3   The first stage decomposition in MRA

The goal is to prove that under the proper condition (3.1), $d_n^m = \text{DPWT}(m,n)$. From Figure 3.3,

$$\tilde{d}_n^1 = \sqrt{2}\sum \overline{h}(l)\, s(n-l) \tag{3.2}$$

After subsampling and renumbering,

$$d_n^1 = \tilde{d}_{2n+p-1}^1 = \sqrt{2}\sum \overline{h}(l)\, s(2n+p-1-l) \tag{3.3}$$

where the delay $p-1$ is the filter order. Using (3.1) in (3.3) results in

$$d_n^1 = \sqrt{2}\sum \overline{h}(l)\int s(t)\,\phi(t-2n-p+1+l)\,dt \tag{3.4}$$

By definition (2.20), the DPWT$(1,n)$ of $s(t)$, with respect to a wavelet $\psi(t)$, is

$$\text{DPWT}(1,n)=\int s(t)\,\psi_{1n}(t)\,dt \tag{3.5}$$

where from (2.33)

$$\psi_{1n}(t) = \frac{1}{\sqrt{2}} \psi\left(\frac{t}{2} - n\right) \tag{3.6}$$

and

$$\psi_{00}(t) = \psi(t) \tag{3.7}$$

The condition for $d_n^1 = \text{DPWT}(1,n)$ comes from equating (3.4) and (3.5), which is

$$\frac{1}{\sqrt{2}} \psi\left(\frac{t}{2} - n\right) = \sqrt{2} \sum \overline{h}(l) \phi(t - 2n - p + 1 + l) \tag{3.8}$$

Letting $t/2 - n = t$ gives

$$\psi(t) = 2 \sum \overline{h}(l) \phi(2t - p + 1 + l) \tag{3.9}$$

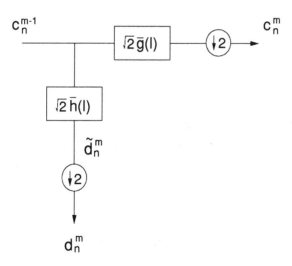

Figure 3.4 The mth stage reconstruction in MRA

Next,

$$c_n^1 = \sqrt{2} \sum \overline{g}(i)\, s(2n + p - 1 - i) \tag{3.10}$$

so that for $m = 2$ in Figure 3.4

$$d_n^2 = 2 \sum_l \sum_i \overline{g}(i)\, \overline{h}(l)\, s(4n + 3p - 3 - 2l - i) \tag{3.11}$$

Since

$$\mathrm{DPWT}(2, n) = \frac{1}{2} \int s(t)\, \psi\left(\frac{t}{4} - n\right) dt \tag{3.12}$$

equating (3.11) to (3.12) and upon using (3.1) yields

$$\frac{1}{2} \psi\left(\frac{t}{4} - n\right) = 2 \sum_l \sum_i \overline{g}(i)\, \overline{h}(l)\, \phi(t - 4n - 3p + 3 + 2l + i) \tag{3.13}$$

Upon using (3.9), this becomes

$$\sum \overline{h}(l)\, \phi\left(\frac{t}{2} - 2n - p + 1 + l\right) = 2 \sum_l \sum_i \overline{g}(i)\, \overline{h}(l)$$

$$\phi(t - 4n - 3p + 3 + 2l + i) \tag{3.14}$$

With $t/2 - 2n - p + 1 + l = t$, this simplifies to

$$\phi(t) = 2 \sum_i \overline{g}(i)\, \phi(2t - p + 1 + i) \tag{3.15}$$

Equations (3.9) and (3.15) are two-scale equations that define the function $\phi(t)$ and $\psi(t)$, via the decomposition filter coefficients. For perfect reconstruction, the synthesis filter coefficients $g(l)$, $h(l)$ must satisfy the relationship (see Section 3.4)

$$\overline{g}(l) = g(p - 1 - l) \tag{3.16}$$

$$\overline{h}(l) = h(p - 1 - l) \tag{3.17}$$

Putting these into (3.9) and (3.15) yields the more familiar two-scale equations

$$\phi(t) = 2 \sum g(l) \, \phi(2t - l) \tag{3.18}$$

and

$$\psi(t) = 2 \sum h(l) \, \phi(2t - l) \tag{3.19}$$

The function $\phi(t)$ is the scaling function associated with the lowpass filter $g(l)$ (or $\overline{g}(l)$), so called because it serves to time-scale the $s(n)$ sequence. The two-scale difference equations are fundamental in the generation of orthonormal, compact support, discrete parameter wavelets. They show how the digital filters $g(l)$ and $h(l)$ determine wholly the scaling function $\phi(t)$ and wavelet $\psi(t)$, not the reverse.

The preceding has shown that $d_n^m = \text{DPWT}(m, n)$ for $m = 1, 2$, when (3.1) holds. Next, it is proved by induction that the result is true for all m. First assume $d_n^m = \text{DPWT}(m, n)$ so that

$$d_n^m = \int s(t) \, \psi_{mn}(t) \, dt \tag{3.20}$$

or

$$d_n^m = \sqrt{2} \sum \overline{h}(l) \, c_{2n+p-1-l}^{m-1} \tag{3.21}$$

From (3.9)

$$\psi_{mn}(t) = 2^{-\frac{(m-2)}{2}} \sum \overline{h}(l) \, \phi\left(\frac{t}{2^{m-1}} - p + 1 + l - 2n\right) \tag{3.22}$$

Substituting (3.22) into (3.20) and equating to (3.21) produces

$$c_{2n+p-1-l}^{m-1} = 2^{-\frac{(m-1)}{2}} \int s(t) \, \phi\left(\frac{t}{2^{m-1}} - p + 1 + l - 2n\right) dt \tag{3.23}$$

Letting $2n + p - 1 - l = n$ gives

$$c_n^{m-1} = \int s(t) \, \phi_{m-1,n}(t) \, dt \tag{3.24}$$

where, similar to (2.33)

$$\phi_{mn}(t) = 2^{-\frac{m}{2}} \phi\left(\frac{t}{2^m} - n\right) \tag{3.25}$$

There is an interesting parallel between (3.20) and (3.24): d_n^m is the inner product of $s(t)$ and $\psi_{mn}(t)$, while c_n^m is the inner product of $s(t)$ and $\phi_{mn}(t)$. With

$$c_n^m = \sqrt{2} \sum \overline{g}(i) \, c_{2n+p-1-i}^{m-1} \tag{3.26}$$

it follows that

$$d_n^{m+1} = \sqrt{2} \sum \overline{h}(l) \, c_{2n+p-1-l}^m \tag{3.27}$$

or from Figure 3.4,

$$d_n^{m+1} = 2 \sum \overline{h}(l) \sum \overline{g}(i) \, c_{4n+3(p-1)-2l-i}^{m-1} \tag{3.28}$$

which becomes, on using (3.23)

$$d_n^{m+1} = 2^{-\frac{(m-1)}{2}} \sum \overline{h}(l) \int s(t)$$

$$2 \sum \overline{g}(i) \, \phi\left(\frac{t}{2^{m-1}} - 3(p-1) + 2l + i - 4n\right) dt \tag{3.29}$$

This simplifies, by using (3.15), to

$$d_n^{m+1} = 2^{-\frac{(m-1)}{2}} \sum \overline{h}(l) \int s(t) \, \phi\left(\frac{t}{2^m} - (p-1) + l - 2n\right) dt \tag{3.30}$$

so that from (3.9)

$$d_n^{m+1} = 2^{-\frac{(m+1)}{2}} \int s(t) \, \psi\left(\frac{t}{2^{m+1}} - n\right) dt = \text{DPWT}(m+1, n) \tag{3.31}$$

Thus the $(m+1)$th result follows from the mth result and since equality already holds for $m = 1, 2$, $d_n^m = \text{DPWT}(m, n)$ holds for any m.

Let

$$\Phi(\omega) = \int \phi(t) \, e^{-j\omega t} \, dt \tag{3.32}$$

$$G(\omega) = \Sigma \, g(l) \, e^{-j\omega l} \tag{3.33}$$

$$H(\omega) = \Sigma \, h(l) \, e^{-j\omega l} \tag{3.34}$$

$$\Psi(\omega) = \int \psi(t) \, e^{-j\omega t} \, dt \tag{3.35}$$

Then the Fourier transformation of (3.18) and (3.19) yields

$$\Phi(\omega) = G\left(\frac{\omega}{2}\right) \Phi\left(\frac{\omega}{2}\right) \tag{3.36}$$

and

$$\Psi(\omega) = H\left(\frac{\omega}{2}\right) \Phi\left(\frac{\omega}{2}\right) \tag{3.37}$$

Iterating (3.36) to infinity results in

$$\Phi(\omega) = \prod_{k=1}^{\infty} G\left(\frac{\omega}{2^k}\right) \Phi(0) \tag{3.38}$$

and

$$\Psi(\omega) = H\left(\frac{\omega}{2}\right) \prod_{k=1}^{\infty} G\left(\frac{\omega}{2^{k+1}}\right) \Phi(0) \tag{3.39}$$

Normalizing the scaling function by

$$\int \phi(t) \, dt = 1 \tag{3.40}$$

leads to

$$\Phi(0) = 1 \tag{3.41}$$

and because of (3.38)

$$G(0) = 1 \tag{3.42}$$

and from (2.16) and (3.37)

$$H(0) = 0 \tag{3.43}$$

Note that MRA has $s(n)$ as input and deals strictly in discrete time. Yet it also implements the DPWT, where both $s(t)$ and $\psi(t)$ are in continuous time, when (3.1) holds. In practice, $s(n)$ generally comes from a direct sampling of $s(t)$, already a lowpass signal. Then (3.1) is not valid and $d_n^m \neq \text{DPWT}(m,n)$ of $s(t)$. Instead, the d_n^m equal the $\text{DPWT}(m,n)$ of the function

$$\tilde{s}(t) = \sum_n s(n)\,\phi(t-n) \tag{3.44}$$

This follows for orthonormal $\phi(t-n)$, for then

$$\int \tilde{s}(t)\,\phi(t-m)\,dt = \sum s(n)\int \phi(t-n)\,\phi(t-m)\,dt \tag{3.45}$$

so that

$$s(m) = \int \tilde{s}(t)\,\phi(t-m)\,dt \tag{3.46}$$

showing that the $s(n)$ are samples of the inner product of $\tilde{s}(t)$ and the scaling function $\phi(t-n)$. When the $s(n)$ do not come from (3.1), and they are generally not, then the details d_n^m are only approximations to the $\text{DPWT}(m,n)$ of $s(t)$, with respect to the wavelet $\psi(t)$. It is possible to improve on the approximation accuracy by prefiltering the $s(n)$. Details are in [34] and [35].

There is one more point to note regarding the MRA computation of the $\text{DPWT}(m,n)$ of $s(t)$. When $s(n)$ comes from (3.1), $d_n^m = \text{DPWT}(m,n)$ of $s(t)$. But the development following (3.44) claims that $d_n^m = \text{DPWT}(m,n)$ of $\tilde{s}(t)$ as well. Since $s(t) \neq \tilde{s}(t)$, why are their $\text{DPWT}(m,n)$ equal? The answer is that they are equal only for $m = 1, 2, \ldots, \infty$, but not for negative m. the projection of $s(t)$ onto $\phi(t-n)$ is $\tilde{s}(t)$ so that $\text{DPWT}(m,n)$ of $s(t)$

and $\tilde{s}(t)$ are equal for m positive.   However, for negative m, the DPWT(m,n) of $s(t)$ contains components of $s(t)$ not in the subspace of $\phi(t-n)$, and hence they will be different from the DPWT(m,n) of $\tilde{s}(t)$.

3.3 Principles of MRA

Multiresolution analysis [19] is the decomposition of a signal $s(t)$ into components of different scales (frequency) of 2^{-m}, m integer.   Associating with each scale (frequency band) is a subspace $\mathbf{V}_m$.   These subspaces are time functions which satisfy the following conditions.

(i)   Containment:

$$\{0\} \rightarrow \ldots \mathbf{V}_2 \subset \mathbf{V}_1 \subset \mathbf{V}_0 \subset \mathbf{V}_{-1} \subset \mathbf{V}_{-2} \ldots \rightarrow \mathbf{L}^2(\mathbf{R}) \qquad (3.47)$$

The subspaces begin with the null space and expands in scales of two to reach the space of all square integrable function.   If a function $s(t)$ is in $\mathbf{V}_i$, then $s(2t)$ is in $\mathbf{V}_{i-1}$, and vice versa.   For example, let $\mathbf{V}_m$ be the subspace of all piecewise constant functions, i.e.,

$$\mathbf{V}_m = \{ s(t) \in \mathbf{L}^2(\mathbf{R}) : s(t) \text{ constant between } 2^m n , 2^m (n+1) \} \quad (3.48)$$

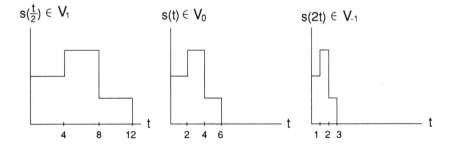

Figure 3.5   Piecewise constant $s(t)$

A typical $s(t)$ is in Figure 3.5, showing that $\mathbf{V}_{-1}$ contains $\mathbf{V}_0$, which contains $\mathbf{V}_1$. Any $s(t)$ in $\mathbf{V}_m$ that is constant between $2^m n$, $2^m(n+1)$ will also have $s(2t)$ constant between $2^{m-1}n$, $2^{m-1}(n+1)$. In the frequency domain, the containment of subspace $\mathbf{V}_0$ by subspace $\mathbf{V}_{-1}$ implies that the spectrum of $\mathbf{V}_0$ is inside the spectrum of $\mathbf{V}_{-1}$, as seen in Figure 3.7.

(ii) Existence of orthonormal scaling functions:
There exists a scaling function $\phi(t) \in \mathbf{V}_0$ such that the set

$$\left\{ \phi_{mn}(t) = 2^{-\frac{m}{2}} \phi(2^{-m}t - n) : n = integer \right\} \tag{3.49}$$

is an orthonormal basis that spans $\mathbf{V}_m$. For example, again let $\mathbf{V}_m$ be the subspace of piecewise constant functions, then

$$\phi(t) = \begin{cases} 1 & , \quad 0 \le t \le 1 \\ 0 & , \quad otherwise \end{cases} \tag{3.50}$$

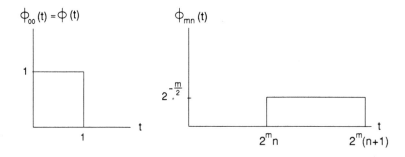

Figure 3.6   Orthonormal basis functions

is a scaling function. Graphically, it is easy to see that the $\phi_{mn}(t)$ of Figure 3.6 form an orthonormal basis in $\mathbf{V}_m$, so that

$$\int \phi_{mn}(t) \phi_{mk}(t) = \delta_{n-k} \tag{3.51}$$

Any $s(t) \in \mathbf{V}_m$ in (3.48) can be exactly represented as a linear combination of $\phi_{mn}(t)$.

(iii) Basis functions defined by two-scale difference equations:
Since $\phi_{0n}(t)$ spans $\mathbf{V}_0$ and $\phi_{-1n}(t)$ spans $\mathbf{V}_{-1}$ and $\mathbf{V}_{-1}$ contains $\mathbf{V}_0$, $\phi_{00}(t) = \phi(t)$ is a linear combination of $\phi_{-1n}(t) = \sqrt{2}\, \phi(2t - n)$, i.e.,

$$\phi(t) = 2 \sum_{l=0}^{p-1} g(l) \phi(2t - l) \tag{3.52}$$

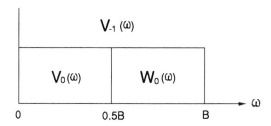

Figure 3.7 Spectra of subspaces

which is a two-scale difference equation, with $g(l)$ the coefficients of combination. Next, consider the spectra $V_{-1}(\omega)$ and $V_0(\omega)$ in Figure 3.7. The bandwidth of $V_0(\omega)$ is one half of that of $V_{-1}(\omega)$ ($\mathbf{V}_{-1}$ contains $\mathbf{V}_0$). Let the band in $V_{-1}(\omega)$ between $0.5B$ and B be $W_0(\omega)$, called the orthogonal complement of $V_0(\omega)$. The subspaces $\mathbf{V}_0$ and $\mathbf{W}_0$ are orthogonal because they have no common frequency components and they are complementary to each other in forming $V_{-1}(\omega)$. Now since $\mathbf{W}_o \subset \mathbf{V}_{-1}$, then a function $\psi(t)$, called the wavelet function, whose translates $\psi(t-n)$ span $\mathbf{W}_0$, can also be written as a linear combination of $\phi_{-1n}(t)$, which span $\mathbf{V}_{-1}$. Thus

$$\psi(t) = 2 \sum_{l=0}^{p-1} h(l) \phi(2t - l) \tag{3.53}$$

which is again a two-scale equation with the coefficients $h(l)$. Moreover, it follows from the inner product preserving property (2.51) that

$$< \phi_{mn}(t), \psi_{mk}(t) >=0 \tag{3.54}$$

for any m, n, k.

(iv) Existence of orthonormal wavelet functions:
Substituting (3.53) into

$$\psi_{mn}(t) = 2^{-\frac{m}{2}} \psi(2^{-m}t - n) \tag{3.55}$$

yields

$$\psi_{mn}(t) = 2^{1-\frac{m}{2}} \sum_l h(l)\, \phi(2^{1-m}t - 2n - l)$$

$$= \sqrt{2} \sum_l h(l)\, \phi_{m-1, 2n+l}(t) \tag{3.56}$$

In view of the orthonormality of $\phi_{mn}(t)$, the $\psi_{mn}(t)$ are also orthonormal provided that

$$2\sum_l h^2(l) = 1 \text{ and } \sum_l h(l) h(l+2k) = \begin{cases} 0 \;, & k=1,2,\ldots \\ \dfrac{1}{2} \;, & k=0 \end{cases} \tag{3.57}$$

With

$$< \psi_{mn}(t), \psi_{jk}(t) >= \delta_{m-j}\,\delta_{n-k} \tag{3.58}$$

the $\psi_{mn}(t)$ generate orthogonal subspaces W_m.

To illustrate the above, consider the $\phi(t)$ of (3.50). Obviously

$$\phi(t) = \phi(2t) + \phi(2t - 1) \tag{3.59}$$

Comparing (3.59) with (3.52) gives

$$g(0) = \frac{1}{2} \quad , \quad g(1) = \frac{1}{2} \tag{3.60}$$

Using (3.52) and (3.53) in (3.54) results in

$$2 \, [\, h(0) \, g(0) + h(1) \, g(1) \,] = 0 \tag{3.61}$$

One choice of $h(l)$ to satisfy (3.61) is

$$h(0) = g(1) = \frac{1}{2} \quad , \quad h(1) = -g(0) = \frac{-1}{2} \tag{3.62}$$

so that

$$\psi(t) = \phi(2t) - \phi(2t - 1) \tag{3.63}$$

which is the Haar wavelet of (2.36).

In (3.52) and (3.53), the coefficients in the two-scale equations are $g(l)$ and $h(l)$. They also appear in (3.18) and (3.19), but as the coefficients of the lowpass and highpass FIR synthesis filters (Figure 3.2). Are they really the same? The following will show that they are. It will also show that the decomposition filters' (Figure 3.1) coefficients $\overline{g}(l) = g(p-1-l)$ and $\overline{h}(l) = h(p-1-l)$.

Given any $s(t) \in L^2(\mathbf{R})$, the MRA of $s(t)$, starting from subspace $\mathbf{V}_0$, first computes the projections of $s(t)$ onto the basis functions $\phi_{0n}(t)$ that span $\mathbf{V}_0$. These are

$$s(n) = \int s(t) \, \phi(t - n) \, dt = \, < s(t) , \phi_{0n}(t) > \tag{3.64}$$

which are identical to (3.1). Then the projection of $s(t)$ onto the space $\mathbf{V}_0$ is

$$\bar{s}(t) = \sum s(n) \, \phi(t - n) \tag{3.65}$$

which is iidentical to (3.44). Figure 3.8 is an illustration of the projection of an $s(t)$ onto the $\mathbf{V}_0$ of Haar functions (Section 3.5). Note that $\bar{s}(t) \in \mathbf{V}_0$ but $s(t) \notin \mathbf{V}_0$, hence $\bar{s}(t) \neq s(t)$.

Now V_0 contains V_1 and W_1 which are orthogonal complements of each other, spanned respectively by $\phi_{1n}(t)$ and $\psi_{1n}(t)$ (see MRA conditions (ii) and (iv)), which are also orthogonal (3.54). Hence another representation of $\tilde{s}(t)$ is

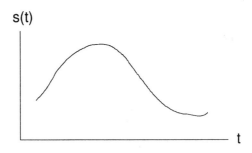

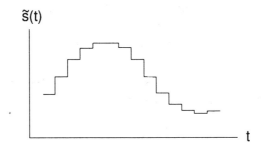

Figure 3.8   Projection of $s(t)$ onto V_0 of Haar functions

$$\tilde{s}(t) = \sum_{n=0}^{M-1} c_n^1 \phi_{1n}(t) + \sum_{n=0}^{M-1} d_n^1 \psi_{1n}(t) \tag{3.66}$$

where

$$c_n^1 = \, <\tilde{s}(t), \phi_{1n}(t)> \tag{3.67}$$

and

$$d_n^1 = <\tilde{s}(t), \psi_{1n}(t)> \tag{3.68}$$

and M not necessarily equal N. Indeed $M = N/2$ because of decimation by two. Putting (3.64) into (3.67) and (3.68) gives

$$c_n^1 = \sum_{k=0}^{N-1} s(k) <\phi_{0k}(t), \phi_{1n}(t)> \tag{3.69}$$

and

$$d_n^1 = \sum_{k=0}^{N-1} s(k) <\phi_{0k}(t), \psi_{1n}(t)> \tag{3.70}$$

Evaluating the inner products and using (3.52) yields

$$<\phi_{0k}(t), \phi_{1n}(t)> = \sqrt{2} \int \phi(t-k) \sum_{l=0}^{p-1} g(l) \phi(t-2n-l) \, dt \tag{3.71}$$

which is non-zero only when $k = 2n + l$ due to (3.51). Therefore

$$c_n^1 = \sqrt{2} \sum_{k=2n}^{2n+p-1} s(k) g(k-2n) \tag{3.72}$$

and similarly

$$d_n^1 = \sqrt{2} \sum_{k=2n}^{2n+p-1} s(k) h(k-2n) \tag{3.73}$$

The summation limits in (3.72) and (3.73) come from the conditions that $g(l)$, $h(l)$ are non-zero only for $0 \le l \le p-1$. Let $k - 2n = p - 1 - l$ in (3.72) so that

$$c_n^1 = \sqrt{2} \sum_{l=0}^{p-1} g(p-1-l) s(2n+p-1-l) \tag{3.74}$$

If the lowpass FIR filter in Figure 3.1 has coefficients

$$\bar{g}(l) = g(p-1-l) \tag{3.75}$$

and in the highpass case

$$\overline{h}(l) = h(p - 1 - l) \tag{3.76}$$

then

$$c_n^1 = \sqrt{2} \sum_{l=0}^{p-1} \overline{g}(l) s(2n + p - 1 - l) \tag{3.77}$$

and

$$d_n^1 = \sqrt{2} \sum_{l=0}^{p-1} \overline{h}(l) s(2n + p - 1 - l) \tag{3.78}$$

In the two equations above, there is a delay of $p - 1$ samples in the output c_n^1 and d_n^1. There is no output c_n^1 until $s(2n + p - 1)$ has arrived. This delay is a consequence of the derivation from (3.74) and is needed for perfect reconstruction (see (3.93)). The first stage of MRA decomposition is in Figure 3.3. The sequence c_n^1 (d_n^1) comes from first filtering $s(n)$ by the filter $\overline{g}(l)$ ($\overline{h}(l)$) and then taking every other sample of its output. By design, $\overline{g}(l)$ ($\overline{h}(l)$) is lowpass (highpass) of bandwidth equal to the lower (upper) half of $|S(\omega)|$. There is no loss of information by downsampling by two the output of $\overline{g}(l)$. The same applies to the output of $\overline{h}(l)$ because of the bandpass sampling theorem [20]. It states that a signal bandlimited between $\omega_2 - \omega_1 = B$ can be sampled at $2B$ without aliasing, even though ω_2 and ω_1 are greater than B. The first stage of the MRA decomposition produces a lowpass component c_n^1, referred to as the smoothed or approximation version of $s(n)$, since its resolution is half of $s(n)$. The other component d_n^1 is the detail or difference, it contains the high frequency details of $s(n)$ that are not in c_n^1.

The next step decomposes c_n^1 into its approximation c_n^2 and detail d_n^2, following identical steps as previous. Thus

$$\sum_n c_n^1 \phi_{1n}(t) = \sum_n c_n^2 \phi_{2n}(t) + \sum_n d_n^2 \psi_{2n}(t) \tag{3.79}$$

with

$$c_n^2 = < \sum_k c_k^1 \phi_{1k}(t), \phi_{2n}(t) > \qquad (3.80)$$

$$= \sum_k c_k^1 < \phi_{1k}(t), \phi_{2n}(t) > \qquad (3.81)$$

$$= \sqrt{2} \sum_k c_k^1 g(k - 2n) \qquad (3.82)$$

$$= \sqrt{2} \sum_{l=0}^{p-1} \overline{g}(l) c_{2n-l+p-1}^1 \qquad (3.83)$$

Similarly

$$d_n^1 = \sqrt{2} \sum_{l=0}^{p-1} \overline{h}(l) c_{2n-l+p-1}^1 \qquad (3.84)$$

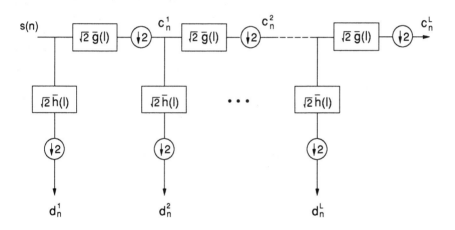

Figure 3.9 MRA decomposition

The filters in the second stage are identical to those in the first. The decomposition continues until it reaches c_N^L, conceivably the dc component of $s(t)$. The block diagram is in Figure 3.9. It is the Mallat [19] algorithm for MRA, also known as the fast wavelet transform [21], when attached with a wavelet transform interpretation, which of course is not necessary. It does, however, implement the DPWT on $s(t)$ with a wavelet $\psi(t)$ (3.19), generated from $\phi(t)$ (3.18), and $d_n^m = \text{DPWT}(m,n)$ provided (3.1) holds. The discussion at the end of Section 3.2 also applies here regarding the equivalence of the $\text{DPWT}(m,n)$ of $s(t)$ and $\tilde{s}(t)$, for positive m only. The algorithm has complexity $O(N)$ and is therefore faster than FFT. The complexity is independent of the number of stages of decomposition. Since a subsampling by two follows each stage, the total number of computations is

$$T = T_o + \frac{T_o}{2} + \frac{T_o}{4} + \ldots = 2\,T_o \tag{3.85}$$

where T_o is the number of computations for the first stage.

One application of MRA is in speech or image compression. By resolving $s(t)$ into components of different resolutions, a coding scheme can assign more bits to represent the lower detail sequences, which have shorter lengths due to successive subsamplings, and less to the higher detail sequences, in order to achieve compression. The receiver will reconstruct $s(t)$ from the coded sequences. A reasonable specification on the receiver would be that if there were no coding, it should give perfect reconstruction, albeit with a permissible delay. The following will show that for perfect reconstruction, the synthesis filters need to have coefficients identical to $g(l)$ and $h(l)$ of (3.52) and (3.53). From (3.64) and (3.66)

$$\sum_{n=0}^{N-1} s(n)\,\phi_{0n}(t) = \sum_{k=0}^{N/2-1} c_k^1\,\phi_{1k}(t) + \sum_{k=0}^{N/2-1} d_k^1\,\psi_{1k}(t) \tag{3.86}$$

so that the reconstruction of $s(n)$ from c_n^1 and d_n^1 is

$$s(n) = \sum_{k=0}^{N/2-1} c_k^1 < \phi_{0n}(t), \phi_{1k}(t) > + \sum_{k=0}^{N/2-1} d_k^1 < \phi_{0n}(t), \psi_{1k}(t) > \tag{3.87}$$

Evaluating the inner products and following the steps that go from (3.69) to (3.73) produces

$$s(n) = \sqrt{2} \sum_{k=n/2}^{(n-p+1)/2} c_k^1\, g(n-2k) + \sqrt{2} \sum_{k=n/2}^{(n-p+1)/2} d_k^1\, h(n-2k) \tag{3.88}$$

Let $l = n - 2k$, then

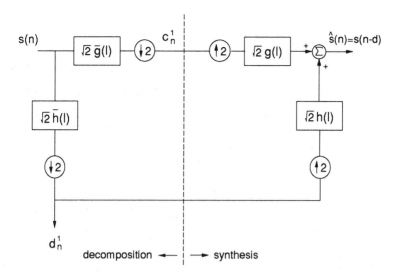

Figure 3.10 Single stage decomposition-synthesis for perfect reconstruction

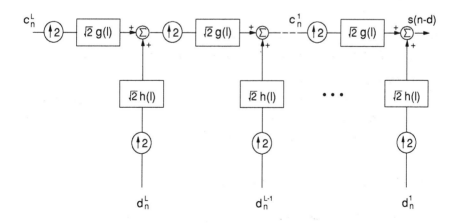

Figure 3.11 Multi-stage perfect reconstruction

$$s(n) = \sqrt{2} \sum_{l=0}^{P-1} g(l) \, c^1_{(n-l)/2} + \sqrt{2} \sum_{l=0}^{P-1} h(l) \, d^1_{(n-l)/2} \tag{3.89}$$

where $c^1_{(n-l)/2} = d^1_{(n-l)/2} = 0$ whenever $n-l$ is odd. Thus perfect reconstruction consists of passing the upsampled c^1_n and d^1_n through anti-aliasing filters $g(l)$ and $h(l)$ and summing the outputs. Figure 3.10 illustrates the simple single stage decomposition-synthesis operations while Figure 3.11 gives the structure for multi-stage synthesis.

This section ends with a summary of MRA:

(i)   It decomposes $s(t)$ into the detail components d^m_n, $m = 1, 2, \ldots, L$ and the coarsest approximation c^L_n via stages of identical lowpass filters $\overline{g}(l)$ and highpass filters $\overline{h}(l)$ and subsampling the outputs.

(ii) The subsampling by two produces a time compression by two and identical $\overline{g}(l)$ ($\overline{h}(l)$) will have bandwidths of the lower (upper) half of the previous $\overline{g}(l)$.

(iii) The details d^m_n are the DPWT coefficients of $s(t)$, i.e., $d^m_n = \text{DPWT}(m, n)$ if (3.1) holds. The wavelet function is generated from (3.53) and scaling function from (3.52). The choice of $g(l)$ and $h(l)$ determines $\phi(t)$ and $\psi(t)$. These functions, however, do not have a role in MRA and it is certainly possible to perform MRA without any reference to $\phi(t)$ or $\psi(t)$.

(iv) The $\psi_{mn}(t)$ from (3.53) form a compact support orthonormal basis if the $g(l)$, $h(l)$ satisfy certain conditions (see the following sections).

The filters in MRA serve to divide a signal's spectrum successively by two. At the same time, their coefficients appear in two-scale equations for generating $\phi(t)$ and $\psi(t)$ to span $\mathbf{V_0}$ and $\mathbf{W_0}$. The properties and design of these filters are the next subjects of discussion.

3.4 Filters for Perfect Reconstruction

This section develops the specifications on the filters $g(l)$, $h(l)$ for perfect reconstruction (PR). Using the Z-domain relationship in Figure 3.12, the Z transform of $\hat{s}(n)$ in Figure 3.10 is

$$\hat{S}(z) = T(z) S(z) + E(z) S(-z) \tag{3.90}$$

with

$$T(z) = \overline{G}(z) G(z) + \overline{H}(z) H(z) \tag{3.91}$$

and

$$E(z) = \overline{G}(-z) G(z) + \overline{H}(-z) H(z) \tag{3.92}$$

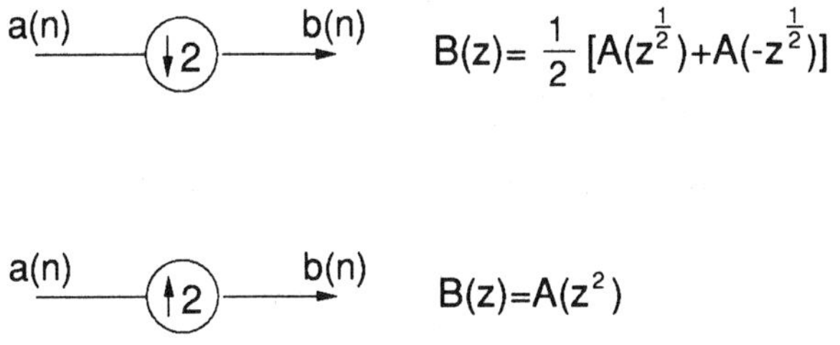

Figure 3.12 Z-domain relationship for down and up sampling [2]

The conditions for PR are

$$T(z) = z^{-d} \tag{3.93}$$

$$E(z) = 0 \tag{3.94}$$

where d is any integer delay. To satisfy (3.94), let

$$G(z) = -\overline{H}(-z) \quad , \quad H(z) = \overline{G}(-z) \tag{3.95}$$

then

$$T(z) = G(z) H(-z) - G(-z) H(z) \tag{3.96}$$

To satisfy (3.93), it is sufficient, but not necessary, to let

$$H(z) = -z^{-(p-1)} G(-z^{-1}) \tag{3.97}$$

This is the paraunitary (lossless) solution [2] where $H(z)$ and $G(z)$ are paraunitary FIR filters of order p, an even number. Then

$$T(z) = z^{-(p-1)}[R(z) + R(-z)] \qquad (3.98)$$

with

$$R(z) = G(z)G(z^{-1}) \qquad (3.99)$$

But

$$G(z) = \sum_{l=0}^{p-1} g(l) z^{-l} \qquad (3.100)$$

so that the condition for PR, from (3.98), is

$$1 = R(z) + R(-z) = 2\sum_l g^2(l) + \sum_i \sum_{k \neq i} g(i) g(k) z^{k-i}$$

$$+ \sum_l \sum_{m \neq l} g(l) g(m) z^{m-l} (-1)^{m-l} \qquad (3.101)$$

where the suppressed summation limits are all from 0 to p-1. The last two terms sum to zero if $k - i = m - l = $ odd. Hence

$$1 = 2\sum_l g^2(l) + 2 \sum_{k-i=even, \neq 0} g(i) g(k) z^{k-i} \qquad (3.102)$$

Applying the constraint, similar to (3.57), of

$$\sum_l g^2(l) = \frac{1}{2} \qquad (3.103)$$

the PR requirement is

$$\sum_i g(i) g(i+2k) = \begin{cases} 0, & k = 1, 2, \dots \\ \dfrac{1}{2}, & k = 0 \end{cases} \qquad (3.104)$$

Additionally, it follows from (3.97) that

$$h(l) = (-1)^l g(p-1-l) \qquad (3.105)$$

so that

$$\sum_l h(l)\, g(l) = 0 \tag{3.106}$$

The impulse responses of the paraunitary filters are orthogonal to each other. It follows from (2.51) that their spectra are also orthogonal. It is also clear from (3.95) that this property for $h(l)$, $g(l)$ applies equally to $\bar{h}(l)$, $\bar{g}(l)$. In summary, the PR conditions on the filters are:

from (3.105) $$h(l) = (-1)^l\, g(p-1-l) \tag{3.107}$$

from (3.97) and (3.95) $$\bar{g}(l) = g(p-1-l) \tag{3.108}$$

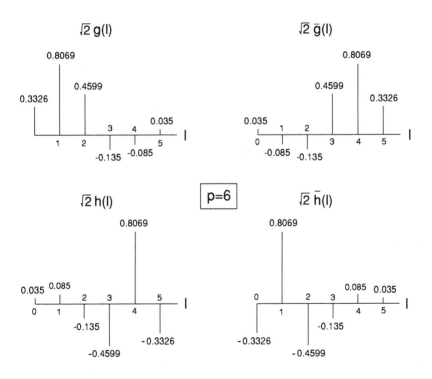

Figure 3.13 PR-paraunitary filter coefficients

from (3.95) $$\overline{h}(l) = (-1)^{l+1} g(l) \qquad (3.109)$$

from (3.109) and (3.105) $$\overline{h}(l) = h(p-1-l) \qquad (3.110)$$

from (3.103) $$\sum_l g^2(l) = \sum_l h^2(l) = \sum_l \overline{g}^2(l) = \sum_l \overline{h}^2(l) = \frac{1}{2} \qquad (3.111)$$

from (3.104) $$\sum_l g(l) g(l+2k) = \begin{cases} 0 & , \quad k=1,2,\ldots \\ \dfrac{1}{2} & , \quad k=0 \end{cases} \qquad (3.112)$$

applies to $h(l), \overline{g}(l), \overline{h}(l)$ as well

Since $g(l)$ determines all the other filters, the whole MRA problem comes down to the design of the filter $g(l)$. Figure 3.13 gives numerical values for the 6th order filters. It is easy to verify (3.107) to (3.112) with this example.

To provide a better understanding of the orthogonality between the lowpass and highpass filters, consider the formulation of the filtering and decimation process in the fast wavelet transform (or MRA) as matrix operations. Let

$$s = [s(0), s(1), \ldots, s(N-1)]^T \qquad (3.113)$$

$$c = \left[c_0^1, c_1^1, \ldots, c_{\frac{N}{2}-1}^1 \right]^T \qquad (3.114)$$

$$d = \left[d_0^1, d_1^1, \ldots, d_{\frac{N}{2}-1}^1 \right]^T \qquad (3.115)$$

and for $p=4$

$$\overline{G} = \sqrt{2} \begin{bmatrix} g(0) & g(1) & g(2) & g(3) & 0 & 0 & . & 0 \\ 0 & 0 & g(0) & g(1) & g(2) & g(3) & . & 0 \\ 0 & 0 & 0 & 0 & g(0) & g(1) & . & 0 \\ \vdots & \vdots & \vdots & \vdots & \vdots & \vdots & \vdots & \vdots \end{bmatrix} \qquad (3.116)$$

which is an $N/2 \times N$ matrix with entries $\overline{G}_{ij} = \sqrt{2} g(-2i+j)$, for example $\overline{G}_{01} = \sqrt{2} g(1)$, and

$$\overline{H} = \sqrt{2} \begin{bmatrix} g(3) & -g(2) & g(1) & -g(0) & 0 & 0 & . & 0 \\ 0 & 0 & g(3) & -g(2) & g(1) & -g(0) & . & 0 \\ 0 & 0 & 0 & 0 & g(3) & -g(2) & . & 0 \\ : & : & : & : & : & : & : & : \end{bmatrix} \quad (3.117)$$

which is an $N/2 \times N$ matrix with entries $\overline{H}_{ij} = \sqrt{2}(-1)^j g(2i+p-1-j)$, then it follows from (3.74) and (3.78) (replacing $\overline{h}(l)$ by (3.109)) that

$$c = \overline{G} S \quad , \quad d = \overline{H} S \quad (3.118)$$

By direct multiplication,

$$\overline{G} \overline{H}^T = 0 \quad (3.119)$$

Hence $\overline{G}$ and $\overline{H}$ are orthogonal by construction. In addition, by direct multiplication and (3.112),

$$\overline{G} \overline{G}^T = I \quad , \quad \overline{H} \overline{H}^T = I \quad (3.120)$$

giving

$$\overline{G}^T \overline{G} \overline{G}^T \overline{G} = \overline{G}^T \overline{G} \quad (3.121)$$

an idempotent matrix which is the projection on the subspace V_1 along W_1. Similarly, $\overline{H}^T \overline{H}$ is the projection on W_1 along V_1. Now by (3.119) and (3.120)

$$(\overline{G} + \overline{H})(\overline{G}^T \overline{G} + \overline{H}^T \overline{H}) = \overline{G} + \overline{H} \quad (3.122)$$

However, $(\overline{G}^T \overline{G} + \overline{H}^T \overline{H}) \neq I$, as easily verified by direct computation. For reconstruction, as in Figure 3.4, there is upsampling by two on the vectors c and d before filtering. Let the upsampling matrix be

$$
U_{N \times \frac{N}{2}} =
\begin{bmatrix}
1 & 0 & 0 & . & . & . & 0 \\
0 & 0 & 0 & . & . & . & 0 \\
0 & 1 & 0 & . & . & . & 0 \\
\vdots & \vdots & \vdots & \vdots & \vdots & \vdots & \vdots \\
0 & 0 & 0 & . & . & . & 1 \\
0 & 0 & 0 & . & . & . & 0
\end{bmatrix}
\tag{3.123}
$$

The reconstruction matrices are therefore ($p = 4$)

$$
G = \sqrt{2}
\begin{bmatrix}
g(3) & g(2) & g(1) & g(0) & 0 & . & . & . & 0 \\
0 & g(3) & g(2) & g(1) & g(0) & . & . & . & 0 \\
0 & 0 & g(3) & g(2) & g(1) & g(0) & . & . & 0 \\
\vdots & \vdots & \vdots & \vdots & \vdots & \vdots & \vdots & \vdots & \vdots \\
0 & 0 & 0 & 0 & 0 & . & . & . & g(3)
\end{bmatrix}_{N \times N} U
$$

$$
= \sqrt{2}
\begin{bmatrix}
g(3) & g(1) & 0 & . & . & . & 0 \\
0 & g(2) & g(0) & . & . & . & 0 \\
0 & g(3) & g(1) & . & . & . & 0 \\
\vdots & \vdots & \vdots & \vdots & \vdots & \vdots & \vdots \\
0 & 0 & 0 & . & . & . & 0
\end{bmatrix}_{N \times \frac{N}{2}}
\tag{3.124}
$$

$$
H = \sqrt{2}
\begin{bmatrix}
-g(0) & g(1) & -g(2) & g(3) & 0 & . & . & . & 0 \\
0 & -g(0) & g(1) & -g(2) & g(3) & . & . & . & 0 \\
0 & 0 & -g(0) & g(1) & -g(2) & g(3) & . & . & 0 \\
\vdots & \vdots & \vdots & \vdots & \vdots & \vdots & \vdots & \vdots & \vdots \\
0 & 0 & 0 & 0 & 0 & . & . & . & -g(0)
\end{bmatrix}_{N \times N} U
$$

$$
= \sqrt{2}
\begin{bmatrix}
-g(0) & -g(2) & 0 & . & . & . & 0 \\
0 & g(1) & g(3) & . & . & . & 0 \\
0 & -g(0) & -g(2) & . & . & . & 0 \\
\vdots & \vdots & \vdots & \vdots & \vdots & \vdots & \vdots \\
0 & 0 & 0 & . & . & . & 0
\end{bmatrix}_{N \times \frac{N}{2}}
\tag{3.125}
$$

The reconstructed output is

$$
\hat{s} = G c + H d
\tag{3.126}
$$

where

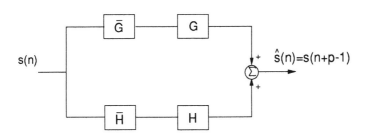

Figure 3.14   MRA in matrix form

$$\hat{s} = [\,\hat{s}(0)\,,\hat{s}(1)\,,\ldots,\hat{s}(N-1)\,]^{T} \qquad (3.127)$$

Substituting (3.118) into (3.126) results in

$$\hat{s} = \begin{bmatrix} O & I \\ O & O \end{bmatrix} s \qquad (3.128)$$

where

$$G\overline{G} + H\overline{H} = \begin{bmatrix} O_{p-1 \times N-p+1} & I_{N-p+1 \times N-p+1} \\ O_{p-1 \times p-1} & O_{N-p+1 \times p-1} \end{bmatrix} \qquad (3.129)$$

which is easy to verify by direct computation. Hence

$$\hat{s}(n) = s(n + p - 1) \qquad (3.130)$$

showing perfect reconstruction but with a delay of p-1. Figure 3.14 is the matrix operation flow chart equivalence of the fast wavelet transform.

The paraunitary filters in (3.97) have amplitude response

$$|H(\omega)| = |G(\pi - \omega)| = |G(\omega + \pi)| \qquad (3.131)$$

Such filters are called quadrature mirror filters (QMF) [22], since they are mirror images of each other with respect to the quadrature frequency $2\pi/4$. A typical though idealized example of their frequency response is in Figure 3.15. With the normalization $G(0) = 1$,

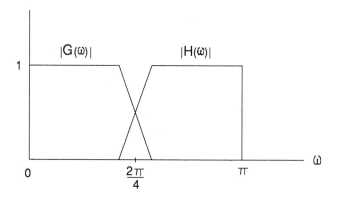

Figure 3.15   Quadrature mirror filters

$$|G(\omega)|^2 + |H(\omega)|^2 = 1 \qquad (3.132)$$

so that the two filters are also power complementary [22]. Combining (3.132) with (3.97) gives the power symmetric condition [22]

$$G(z)G(z^{-1}) + G(-z)G(-z^{-1}) = 1 \qquad (3.133)$$

While the paraunitary filters possess some useful properties and lead to compact support orthonormal wavelets, they cannot be FIR linear phase filters except for the trivial case of $p = 2$, which gives the Haar wavelet [22]. Thus these filters will not find application in image processing as nonlinear phase filters distort an image [1]. To obtain PR, linear phase FIR filters, it is necessary to give up paraunitariness [22]. See also Section 3.7 on the same subject.

3.5 Paraunitary Filters and Orthonormal Wavelets

The previous section has introduced paraunitary filters as one way to achieve PR. As a result, the corresponding wavelets are orthonormal. The proof to follow is by imposing the compact support, orthonormal requirements on $\phi(t)$ and $\psi(t)$ and then show via (3.52) and (3.53) that $g(l)$ and $h(l)$ are necessarily paraunitary.

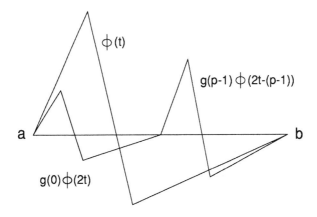

Figure 3.16   Constructing $\phi(t)$ from $\phi(t) = 2 \sum_{l=0}^{p-1} g(l) \phi(2t - l)$

Suppose $g(l) \neq 0$ only for $l = 0, 1, \ldots, p - 1$.   Since $\phi(t)$ is a linear combination (3.52) of the $\phi(2t - l)$, if $\phi(t)$ has support between $t = a$ and

$t = b$, then $\phi(2t - l)$ can also only have support on the same interval. A graphical illustration of this reasoning is in Figure 3.16. Hence $\phi(t)$ is non-zero only if

$$2t - (p - 1)T \leq b \quad \text{and} \quad 2t - 0 \geq a \qquad (3.134)$$

and at the extremes, when $t = a$ and $t = b$,

$$b \leq p - 1 \quad \text{and} \quad a \geq 0 \qquad (3.135)$$

where T is the sampling interval taken to be unity for convenience. The conclusion is if $g(l)$ is supported between 0 and $p - 1$, so is $\phi(t)$. An example is the Haar scaling function, supported between 0 and 1 since there are only $g(0)$ and $g(1)$ to generate $\phi(t)$.

Next, the orthogonality of $\phi_{mn}(t)$ means, for $m = 0$,

$$< \phi_{0l}(t), \phi_{0n}(t) > = \delta(n - l) \qquad (3.136)$$

and on using (2.51)

$$< \Phi(\omega) e^{-j\omega l}, \Phi(\omega) e^{-j\omega n} > = \delta(n - l) \qquad (3.137)$$

Hence for $k = n - l$

$$\frac{1}{2\pi} \int_{-\infty}^{\infty} |\Phi(\omega)|^2 e^{j\omega k} d\omega = \delta(k) \qquad (3.138)$$

which implies $|\Phi(\omega)|^2 = 1$ for all ω, and $\phi(t)$ is an impulse. A better alternative is by recognizing that the autocorrelation of $\phi(t)$ is

$$R_{\phi\phi}(\tau) = \frac{1}{2\pi} \int_{-\infty}^{\infty} |\Phi(\omega)|^2 e^{j\omega\tau} d\omega \qquad (3.139)$$

Sampling it at $\tau = k$, yields, with (3.138),

$$R_{\phi\phi}(k) = \delta(k) \qquad (3.140)$$

Taking the Fourier transform gives

$$\sum_{k=-\infty}^{\infty} |\Phi(\omega + 2\pi k)|^2 = 1 \quad, \quad \text{for all } \omega \qquad (3.141)$$

as a condition for orthonormal $\phi_{0n}(t)$, on $G(\omega)$, via (3.37). An example is the Haar scaling function in Figure 3.17. Several interesting constraints on $\phi(t)$ and the coefficients $g(l)$ follow as a result of (3.141). Since $\Phi(0)=1$ from (3.40), the admissibility condition of (2.16) that $\Psi(0)=0$ forces the condition $H(0)=0$ through (3.37). Hence from (3.132),

$$G(0) = \sum_{l=0}^{p-1} g(l) = 1 \qquad (3.142)$$

Applying $\Phi(0)=1$ to (3.141) gives (see Figure 3.17 for an example)

$$\Phi(2\pi k) = 0 \quad , \qquad k \neq 0 \qquad (3.143)$$

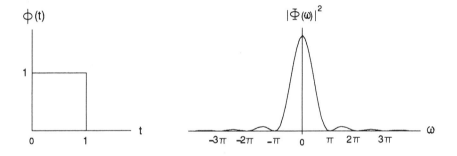

Figure 3.17   The Haar scaling function

From the Poisson Summation formula [3]

$$\sum_{k=-\infty}^{\infty} \phi(t+k) = \sum_{k=-\infty}^{\infty} \Phi(2\pi k)\, e^{j2\pi kt} \qquad (3.144)$$

The condition (3.143) implies

$$\sum_{k=-\infty}^{\infty} \phi(t+k) = 1 \qquad (3.145)$$

and in particular for j integers

$$\sum_{j=-\infty}^{\infty} \phi(j) = 1 \qquad (3.146)$$

Using (3.143) with $\omega = 2\pi$ in (3.36) gives

$$G(\pi) = 0 = \sum_{l=0}^{p-1} (-1)^l g(l) \qquad (3.147)$$

Adding (3.147) to (3.142) gives

$$\sum_{l=0}^{\lfloor (p-1)/2 \rfloor} g(2l) = \frac{1}{2} \qquad (3.148)$$

and because of (3.142)

$$\sum_{l=0}^{\lfloor (p-2)/2 \rfloor} g(2l+1) = \frac{1}{2} \qquad (3.149)$$

Continuing on, orthogonality across n at $m = 1$ requires

$$< \phi_{1l}(t), \phi_{1n}(t) > = < \sqrt{2}\,\Phi(2\omega)\, e^{-j2\omega l}, \sqrt{2}\,\Phi(2\omega)\, e^{-j2\omega n} >$$

$$= \delta(n-l) \qquad (3.150)$$

Following the same steps from (3.136) to (3.141) yields

$$\sum_{k=-\infty}^{\infty} |\Phi(2\omega + 2\pi k)|^2 = 1 \qquad (3.151)$$

Substituting in (3.37), this becomes

$$\sum_{k=-\infty}^{\infty} |G(\omega + \pi k)|^2 |\Phi(\omega + \pi k)|^2 = 1 \qquad (3.152)$$

Now

$$G(\omega + \pi k) = \begin{cases} G(\omega), & k \text{ even} \\ G(\omega + \pi), & k \text{ odd} \end{cases} \qquad (3.153)$$

because $G(\omega)$ is the frequency response of a digital filter at a sampling frequency of 2π. Putting this into (3.152) yields

$$|G(\omega)|^2 \sum_{k=even} |\Phi(\omega+\pi k)|^2 + |G(\omega+\pi)|^2 \sum_{k=odd} |\Phi(\omega+\pi k)|^2 = 1 \quad (3.154)$$

For even k, it follows from (3.141) that

$$\sum_{k=even} |\Phi(\omega+\pi k)|^2 = 1 \quad (3.155)$$

Additionally, since

$$\sum_{k=odd} |\Phi(\omega+\pi k)|^2 = \sum_{k=-\infty}^{\infty} |\Phi(\omega+(2k+1)\pi)|^2$$

$$= \sum_{k=-\infty}^{\infty} |\Phi(\omega+\pi+2\pi k)|^2 = 1 \quad (3.156)$$

condition (3.154) is

$$|G(\omega)|^2 + |G(\omega+\pi)|^2 = 1 \quad (3.157)$$

which is identical to (3.133), the requirement for lowpass PR-QMF.

Next, note from (3.52) and (3.53) that if $\phi(t)$ has support between 0 to $p-1$, so has $\psi(t)$. Then by following the same procedures from (3.136) to (3.157), with (3.37) replacing (3.36), the result

$$|H(\omega)|^2 + |H(\omega+\pi)|^2 = 1 \quad (3.158)$$

is obtained. It is the requirement for highpass PR-QMF.

Lastly, the condition

$$<\phi_{mn}(t), \psi_{kl}(t)> = 0 \quad (3.159)$$

implies

$$\sum_k \Phi(\omega+2\pi k) \Psi^*(\omega+2\pi k) = 0 \quad (3.160)$$

and

$$\sum_k \Phi(2\omega+2\pi k) \Psi^*(2\omega+2\pi k) = 0 \quad (3.161)$$

obtained by following the development that produces (3.141) and (3.151). Taking similar steps as those after (3.151) leads to

$$G(\omega) H^*(\omega) + G(\omega+\pi) H^*(\omega+\pi) = 0 \qquad (3.162)$$

or

$$G(z) H(z^{-1}) + G(-z) H(-z^{-1}) = 0 \qquad (3.163)$$

which the paraunitary filter specification of (3.97) satisfies.

From (3.132), since $G(0) = 1$ (3.142), an important constraint on the wavelet is, from (3.30) and (3.40),

$$\int \psi(t)\, dt = H(0) = \sum_{l=0}^{p-1} h(l) = 0 \qquad (3.164)$$

i.e., it has no dc component, which is the admissibility condition (2.16). In addition, (3.164) is equivalent to

$$H(z = e^{j\omega})|_{\omega=0} = H(z = 1) = 0 \qquad (3.165)$$

i.e. $H(z)$ must have at least one zero at $z = 1$. It then follows from (3.97) that

$$H(z=1) = G(z=-1) = 0 \qquad (3.166)$$

Hence $G(z)$ must have at least one zero at $z = -1$.

Beginning with the compact support, orthonormal specification, this section has shown, via the two-scale equations from MRA, that the requisite filters are PR-QMF. The filter order p, an even number, determines the support width of $\phi(t)$ and $\psi(t)$. An additional and important requirement on $G(z)$ is that it must have at least one zero at $z = -1$. The smoothness of $\psi(t)$ improves with the number of zeros $G(z)$ has at $z = -1$ (or $\omega = \pi$). A measure of smoothness is regularity [18] (see Appendix B).

3.6 Filter Design for Orthonormal Wavelets

The remaining problem is the design of $G(z)$ that obeys all the specifications above. However, since $G(z)$ appears in every stage of MRA, resulting in (3.38), it is important that it possesses special properties to ensure

(i)   convergence of

$$\Phi(\omega) = \prod_{k=1}^{\infty} G\left(\frac{\omega}{2^k}\right)$$

(3.167)

(ii) a continuous $\phi(t)$.

Examples of filters whose $\phi(t)$ and $\psi(t)$ are discontinuous are in [15]. To find these properties, write

$$G(z) = \left(\frac{1+z^{-1}}{2}\right)^{L} A(z)$$

(3.168)

where $1 \le L \le p - 1$ and $A(z)$ is a polynomial in z^{-1} of degree $p - 1 - L$. This factorization is always possible because $G(z)$ must have at least one zero at $z = -1$. Here (3.168) assumes that it has L zeros at $z = -1$. The theorem in Appendix B states that if

(i) $|A(\omega) - 1| < \beta |\omega|^{\alpha}$, β any positive constant and $0 < \alpha \le 1$, then $\Phi(\omega)$ converges and obeys

$$|\Phi(\omega)| < \frac{c}{(1+|\omega|)^{r}}$$

(3.169)

where c is a constant, $r = L - b$, $b = \log_2 B$, and $B = \max_{\omega} |A(\omega)|$.

(ii) $B < 2^{L-1}$, i.e. $L - b = r > 1$, then $\phi(t)$ is a continuous function.

The number $r > 1$, not necessarily an integer, is the regularity of the filter $G(z)$.

With reference to the closing remark in Appendix B, the next step is the construction of a $G(z)$ that satisfies $B < 2^{L-1}$. Putting (3.168) into (3.157) results in

$$\left| \frac{1+e^{-j\omega}}{2} \right|^{2L} |A(\omega)|^2 + \left| \frac{1-e^{-j\omega}}{2} \right|^{2L} |A(\omega+\pi)|^2 = 1 \qquad (3.170)$$

Let

$$x = \sin^2\left(\frac{\omega}{2}\right) = \frac{1-\cos(\omega)}{2} \qquad (3.171)$$

so that

$$1-x = \frac{1+\cos(\omega)}{2} \qquad (3.172)$$

The function $|A(\omega)|^2$ is a polynomial in $\cos(\omega)$. Letting

$$|A(\omega)|^2 = p(x) \quad , \quad |A(\omega+\pi)|^2 = p(1-x) \qquad (3.173)$$

changes (3.170) to

$$(1-x)^L p(x) + x^L p(1-x) = 1 \qquad (3.174)$$

The solution of this equation, with $L = p/2$, is (Appendix C)

$$|A(\omega)|^2 = \sum_{l=0}^{L-1} \binom{L+l-1}{l} \sin\left(\frac{\omega}{2}\right)^{2l} \qquad (3.175)$$

Finding $A(\omega)$ from (3.175) generally requires spectral factorization. Consider instead a specific example that gives the Daubechies [18] wavelet for $p = 4$. Let $L = 2$, then

$$|A(\omega)|^2 = 1 + 2\sin^2\left(\frac{\omega}{2}\right) \qquad (3.176)$$

Since $L = p/2 = 2$,

$$A(\omega) = a(0) + a(1)e^{-j\omega} \qquad (3.177)$$

and

$$|A(\omega)|^2 = a^2(0) + a^2(1) + 2a(0)a(1)$$

$$-4a(0)a(1)\sin^2\left(\frac{\omega}{2}\right) \tag{3.178}$$

Equating (3.176) to (3.178) gives the coefficients

$$a(0) = \frac{1+\sqrt{3}}{2} \quad , \quad a(1) = \frac{1-\sqrt{3}}{2} \tag{3.179}$$

which is the minimum phase solution. The non-minimum phase solution is

$$a(0) = \frac{1-\sqrt{3}}{2} \quad , \quad a(1) = \frac{1+\sqrt{3}}{2} \tag{3.180}$$

Then from (3.168), the filter is

$$G(z) = \frac{1}{8}[1+\sqrt{3}+(3+\sqrt{3})z^{-1}+(3-\sqrt{3})z^{-2}+(1-\sqrt{3})z^{-3}] \tag{3.181}$$

As mentioned at the end of Section 3.4, these FIR, PR-QMF are not linear phase, except for $p=2$. Their coefficients, and correspondingly, the wavelet $\psi(t)$ that these filters generate, are not symmetric nor antisymmetric. The linear phase property is important for certain applications such as image coding and [23] gives a design for infinite impulse response wavelet filters that are linear phase. The wavelets, although still orthonormal, are however necessarily of infinite support.

3.7 Biorthogonal Filters

In [24], the introduction of biorthogonal filters produces PR, FIR linear phase filters but at the expense of having different analysis and synthesis scaling functions and wavelets. From Figure 3.18 and (3.91)-(3.94), the PR conditions are

$$\overline{G}(-z)G(z) + \overline{H}(-z)H(z) = 0 \tag{3.182}$$

and

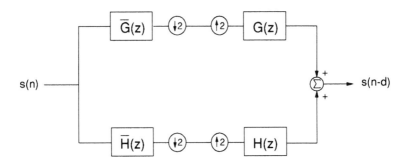

Figure 3.18 PR biorthogonal filters

$$\overline{G}(z)\,G(z)+\overline{H}(z)\,H(z)=z^{-d} \qquad (3.183)$$

To satisfy (3.182), let

$$G(z)=-\overline{H}(-z) \quad , \quad H(z)=\overline{G}(-z) \qquad (3.184)$$

so that

$$G(z)\,\overline{H}(z)=-\overline{H}(z)\,\overline{H}(-z) \qquad (3.185)$$

and

$$\overline{G}(z)\,H(z)=\overline{G}(z)\,\overline{G}(-z) \qquad (3.186)$$

The products on the right hand side of (3.185) and (3.186) must have coefficients for z^l equal to zero, l odd. The same applies to the left hand side, hence for

$$G(z)=\sum_{l=0}^{p-1} g(l)\,z^{-l} \quad , \quad \overline{H}(z)=\sum_{l=0}^{q-1}\overline{h}(l)\,z^{-l} \qquad (3.187)$$

where $G(z)$ and $\overline{H}(z)$ can now have different orders p and q, the following holds

$$\sum_{l=0}^{p-1} g(l)\,[\,\overline{h}(l+k)+\overline{h}(l-k)\,]=0 \quad , \quad k \text{ odd} \qquad (3.188)$$

$$\sum_{l=0}^{p-1} \overline{g}(l)\,[\,h(l+k)+h(l-k)\,]=0 \quad , \qquad k \text{ odd} \qquad (3.189)$$

where $\overline{g}(l)$ and $h(l)$ are similar to $g(l)$ and $\overline{h}(l)$ in (3.187). Further, (3.183) implies that each of the product in the left hand side can only have a non-zero coefficient at z^{-d}. This requirement is

$$\sum_{l=0}^{p-1} \overline{g}(l)\,g(k-l)=\delta(k-d) \qquad (3.190)$$

$$\sum_{l=0}^{q-1} \overline{h}(l)\,h(k-l)=\delta(k-d) \qquad (3.191)$$

Equations (3.188) to (3.191) define a system of biorthogonal filters. Design procedures and numerical examples are in [24]. The filters can be linear phase such that (i) $\overline{G}(z)$ and $\overline{H}(z)$ can be both symmetric of odd order, differing by odd multiples of two, or (ii) $\overline{G}(z)$ or $\overline{H}(z)$ can be symmetric, with the other antisymmetric, with both orders even and are equal or differ by even multiples of two. However, $\overline{g}(l)$ and $g(l)$, and similarly $\overline{h}(l)$ and $h(l)$ no longer obey (3.107) to (3.110). Consequently, the DPWT and inverse DPWT will need different wavelets. Let

$$\phi(t)=2\sum_{l=0}^{p-1} \overline{g}(l)\,\phi(2t-l) \qquad (3.192)$$

and

$$\psi(t)=2\sum_{l=0}^{q-1} \overline{h}(l)\,\phi(2t-l) \qquad (3.193)$$

be the scaling function and wavelet for decomposition, and for synthesis their duals (Sections 2.3 and 1.3)

$$\tilde{\phi}(t)=2\sum_{l=0}^{p-1} g(l)\,\phi(2t-l) \qquad (3.194)$$

and

$$\tilde{\psi}(t)=2\sum_{l=0}^{q-1} h(l)\,\phi(2t-l) \qquad (3.195)$$

The representation for $s(t)$ is (see (2.25))

$$s(t) = \sum_m \sum_n < s(t), \psi_{mn}(t) > \tilde{\psi}_{mn}(t) \tag{3.196}$$

The following orthonormal conditions hold for the scaling and wavelet functions:

$$< \phi(t-k), \tilde{\phi}(t-l) >= \delta(k-l)$$

$$< \psi(t-k), \tilde{\psi}(t-l) >= \delta(k-l)$$

$$< \phi(t-k), \tilde{\psi}(t-l) >= 0$$

$$< \psi(t-k), \tilde{\phi}(t-l) >= 0 \tag{3.197}$$

and in particular (see (3.58))

$$< \psi_{mn}(t), \tilde{\psi}_{jk}(t) >= \delta_{m-j} \, \delta_{n-k} \tag{3.198}$$

3.8 Wavelet Construction from Two-Scale Equations

Although there are wavelets that are explicit functions of time, such as those in (2.4) to (2.7), the wavelets associated with two-scale equations are not. Recursions are necessary to determine them. The wavelet construction equations are:

from (3.18)
$$\phi(t) = 2 \sum_{l=0}^{p-1} g(l) \, \phi(2t-l) \tag{3.199}$$

from (3.38)-(3.41)
$$\Phi(\omega) = \prod_{k=1}^{\infty} G\left(\frac{\omega}{2^k}\right) \tag{3.200}$$

from (3.19)
$$\psi(t) = 2 \sum_{l=0}^{p-1} h(l) \, \phi(2t-l) \tag{3.201}$$

from (3.37)
$$\Psi(\omega) = H\left(\frac{\omega}{2}\right) \Phi\left(\frac{\omega}{2}\right) \tag{3.202}$$

Given $g(l)$ and $h(l)$, the first step is the generation of $\phi(t)$ and then $\psi(t)$ will follow from (3.201). There are three methods of construction [25] but only one is of practical value. Their description below includes some examples that are popular spline functions for curve fitting or interpolation.

M1:

Iterate

$$\phi_{k+1}(t) = 2 \sum_{l=0}^{p-1} g(l) \phi_k(2t-l)$$ (3.203)

for $k = 0, 1, 2, \ldots, \infty,$, using

$$\phi_0(t) = \begin{cases} 1 & , & 0 < t < 1 \\ 0 & , & otherwise \end{cases}$$ (3.204)

Then the scaling function

$$\phi(t) = \lim_{k \to \infty} \phi_{k+1}(t) = \lim_{k \to \infty} \phi_k(t)$$ (3.205)

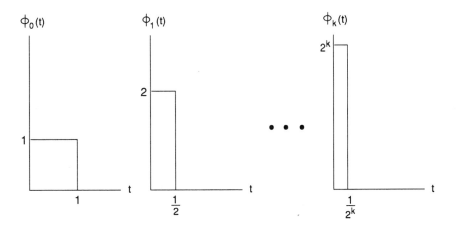

Figure 3.19 $\phi_k(t)$ for Example 3.1

Example 3.1

Let $p = 1$, $g(0) = 1$, then the successive $\phi_k(t)$ are shown in Figure 3.19. As $k \to \infty$, $1/2^k \to 0$, giving $\phi(t)$ as an impulse at $t = 0$. This example shows that $\phi(t)$ is not necessarily a smooth function.

Example 3.2

Let $p = 2$, $g(0) = 1/2$, $g(1) = 1/2$. Starting the iteration (3.203) with $\phi_0(t)$, the gate function, the $\phi_k(t)$ as $t \to \infty$ is still the gate function, which is the scaling function for the Haar wavelet.

Example 3.3

Let $p = 3$, $g(0) = 1/4$, $g(1) = 1/2$, $g(2) = 1/4$. Figure 3.20 displays the successive $\phi_k(t)$. As $k \to \infty$, the incremental steps disappear, resulting in the hat function. Note that this scaling function does not give rise to orthonormal wavelets. Neither is it orthonormal since $\int \phi^2(t)\, dt = 2/3$.

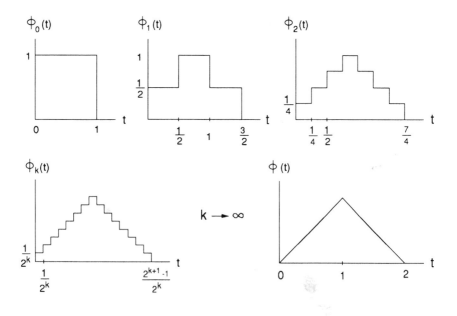

Figure 3.20 $\phi_k(t)$ for Example 3.3

The method **M1** is graphical and is more instructional than practical since it will be difficult to construct the $\phi_k(t)$ when p is of high orders.

M2:

Compute (3.200) and take its inverse Fourier transform to give $\phi(t)$.

Example 3.1 (repeated)

Here $G(\omega) = g(0) = 1$ so that $G(\omega/2^k) = 1$ for all k. Hence $\Phi(\omega) = 1$ and $\phi(t)$ is the impulse function.

Example 3.2 (repeated)

Here

$$G\left(\frac{\omega}{2^k}\right) = \frac{1}{2}\left(1 + e^{-j\omega/2^k}\right) \tag{3.206}$$

Using

$$1 + e^{-j\omega/2^k} = \frac{1 - e^{-j\omega/2^{k-1}}}{1 - e^{-j\omega/2^k}} \tag{3.207}$$

in (3.206) and (3.200) gives

$$\Phi(\omega) = \lim_{m \to \infty} \prod_{k=1}^{m} G\left(\frac{\omega}{2^k}\right) = \lim_{m \to \infty} \left\{ \frac{1 - e^{-j\omega}}{2^m(1 - e^{-j\omega/2^m})} \right\}$$

$$= \frac{1 - e^{-j\omega}}{j\omega} = \int_0^1 \phi(t) e^{-j\omega t}\, dt \tag{3.208}$$

yielding $\phi(t)$ as the gate function.

Example 3.3 (repeated)

Here

$$G\left(\frac{\omega}{2^k}\right) = \frac{1}{4}\left(1 + e^{-j\omega/2^k}\right)^2 \tag{3.209}$$

and applying the results of Example 3.2 gives

$$\Phi(\omega) = \left(\frac{1 - e^{-j\omega}}{j\omega}\right)^2 \tag{3.210}$$

which is the square of (3.208). Hence $\phi(t)$ is the hat function, the convolution of the gate function with itself.

Example 3.4

Let $p = 5$, $g(0) = 1/16$, $g(1) = 4/16$, $g(2) = 6/16$, $g(3) = 4/16$, $g(4) = 1/16$. Then

$$G\left(\frac{\omega}{2^k}\right) = \frac{1}{16}\left(1 + 2e^{-j\omega/2^k} + e^{-j\omega/2^{k-1}}\right)^2 \tag{3.211}$$

which is the square of (3.210). Hence $\phi(t)$ is the cubic B spline, the convolution of the hat function with itself.

The examples in **M2** all possess $\Phi(\omega)$ that have easily recognizable $\phi(t)$; but **M2** will fail if this is not the case.

M3:

Compute (3.199) by recursion. First note from the conclusion after (3.135) that $\phi(t)$ is zero outside $t = 0$ and $t = p - 1$ and since it is a continuous function, it must also be zero at $t = 0$ and $t = p - 1$. Hence for j an integer, $\phi(t)$ is non-zero only for $j \geq 1$ and $j \leq p - 2$. Thus from (3.199), there are $p - 2$ linear equations:

$$\phi(j) = 2\sum_{l=0}^{p-1} g(l)\phi(2j - l) , \qquad j = 1, 2, \ldots, p - 2 \tag{3.212}$$

This gives rise to the matrix equation

$$L\phi = \phi \tag{3.213}$$

where

$$\phi = [\phi(1), \phi(2), \ldots, \phi(p - 2)]^T \tag{3.214}$$

and

$$L_{ik} = 2 g (2i + 1 - k) \tag{3.215}$$

The system in (3.213) is redundant and contains only $p - 3$ independent equations. Discarding one equation, the last one say, in (3.213) and add in the constraint $\sum_j \phi(j) = 1$ (3.146) will give $p - 2$ independent equations to solve for ϕ. Putting these $\phi(j)$ into the recursion

$$\phi\left(\frac{d}{2^k}\right) = 2 \sum_{l=0}^{p-1} g(l) \, \phi\left(\frac{d}{2^{k-1}} - l\right) \tag{3.216}$$

and evaluate for $k = 1, 2, \ldots, k^*$ and $d = 1, 3, 5, \ldots, < 2^k (p - 1)$ will produce $\phi(d/2^k)$ to any time resolution 2^{-k^*}. Then $\psi(d/2^k)$ is similarly obtained via (3.201).

Example 3.5

Using the filter coefficients in (3.181), since $p = 4$, there is only a single independent equation from (3.213). It is

$$\phi(1) = 2 g(0) \, \phi(2) + 2 g(1) \, \phi(1) \tag{3.217}$$

and from (3.146)

$$1 = \phi(1) + \phi(2) \tag{3.218}$$

Solving these two equations gives

$$\phi(1) = \frac{1 + \sqrt{3}}{2} \quad , \quad \phi(2) = \frac{1 - \sqrt{3}}{2} \tag{3.219}$$

Next, let $k = 1$ in (3.216) to give

$$\phi\left(\frac{1}{2}\right) = 2 g(0) \, \phi(1) \tag{3.220}$$

$$\phi\left(\frac{3}{2}\right) = 2 g(1) \, \phi(2) + 2 g(2) \, \phi(1) \tag{3.221}$$

$$\phi\left(\frac{5}{2}\right) = 2 g(3) \, \phi(2) \tag{3.222}$$

Similarly for $k = 2$,

$$\phi\left(\frac{1}{4}\right) = 2\,g\,(0)\,\phi\left(\frac{1}{2}\right) \tag{3.223}$$

$$\phi\left(\frac{3}{4}\right) = 2\,g\,(0)\,\phi\left(\frac{3}{2}\right) + 2\,g\,(1)\,\phi\left(\frac{1}{2}\right) \tag{3.224}$$

This recursion is easy to program and the results are the $\phi(t)$ and $\psi(t)$ in Figure 3.21.

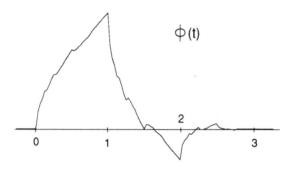

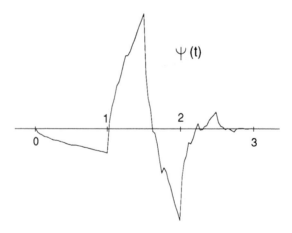

Figure 3.21 $\phi(t)$ and $\psi(t)$ for Example 3.5

3.9 Two Dimensional Wavelets

Since a major potential application of wavelets is in image processing, 2-D wavelet transform is a necessity. The subject, however, is still in an evolving stage [26] and this section will discuss only the extension of 1-D wavelets to the 2-D case.

The idea is to first form a 1-D sequence from the 2-D image row sequences, do a 1-D MRA, restore the MRA outputs to a 2-D format and repeat another MRA to the 1-D column sequences. The two steps of restoring to a 2-D sequence and forming a 1-D column sequence can be combined efficiently by appropriately selecting the proper points directly from the 1-D MRA outputs. As seen in Figure 3.22, after the 1-D row MRA, each lowpass and highpass output goes through a 2-D restoration and 1-D column formation process and then move on to another MRA. Let t_1, t_2 be the 2-D coordinates and $L =$ lowpass, $H =$ highpass. Then the 2-D separable scaling function is

$$\phi^{(1)}(t_1, t_2) = \phi(t_1)\phi(t_2) \quad , \qquad LL \qquad\qquad (3.225)$$

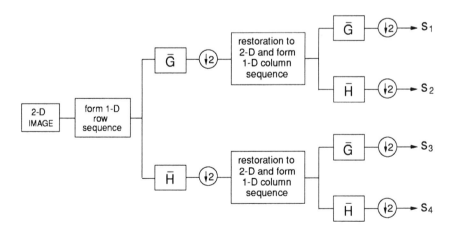

Figure 3.22 2-D MRA decomposition

and the 2-D separable wavelets are

$$\psi^{(2)}(t_1, t_2) = \phi(t_1)\psi(t_2) \quad , \qquad LH \tag{3.226}$$

$$\psi^{(3)}(t_1, t_2) = \psi(t_1)\phi(t_2) \quad , \qquad HL \tag{3.227}$$

$$\psi^{(4)}(t_1, t_2) = \phi(t_1)\phi(t_2) \quad , \qquad HH \tag{3.228}$$

with the corresponding wavelet coefficients s_2, s_3 and s_4.

It is easy to verify that the $\psi^{(i)}$ are orthonormal wavelets, i.e.,

$$\int\int \psi^{(i)}(t_1, t_2)\,dt_1\,dt_2 = 0 \tag{3.229}$$

and

$$<\psi_{mn}^{(i)}, \psi_{kl}^{(i)}> = \delta_{m-k}\,\delta_{n-l} \tag{3.230}$$

etc.

The scheme of separable 2-D processing, while simple and uses available 1-D filters, has disadvantages when compared to a genuine, 2-D MRA with non-separable filters. The latter possesses more freedom in design, can provide a better frequency and even linear phase response, and have non-rectangular sampling [26,27].

Appendix B - Theorem for Convergence of $\Phi(\omega)$

Write

$$G(z) = \left(\frac{1+z^{-1}}{2}\right)^L A(z) \tag{B.1}$$

as a product of its L roots at $z=-1$ and an $p-1-L$ degree polynomial. Since PR-QMF must have at least a zero at $z=-1$ (3.166), $L \geq 1$. Also

$$G(z=1)=1 \tag{B.2}$$

due to (3.133), hence

$$A(z=1)=1 \tag{B.3}$$

Now because of (3.133),

$$|G(z)| \leq 1 \tag{B.4}$$

implying through $(B.1)$ that

$$|A(z)| \leq \gamma \quad , \qquad \text{some positive constant} \tag{B.5}$$

Theorem: For $G(z)$ in $(B.1)$, if $|A(\omega)-1| \leq \beta |\omega|^\alpha$ with β any positive constant and $0 < \alpha \leq 1$, then

(i)
$$\Phi(\omega) = \prod_{k=1}^{\infty} G\left(\frac{\omega}{2^k}\right) \text{ converges} \tag{B.6}$$

(ii)
$$|\Phi(\omega)| < \frac{c}{(1+|\omega|)^{L-b}} \tag{B.7}$$

$$c = \text{any constant} \quad , \qquad B = \max_{\omega} |A(\omega)| \quad , \qquad b = \log_2 B \tag{B.8}$$

(iii) if $B < 2^{L-1}$, then $\phi(t)$ is a continuous function

Proof:

First note that since $A(\omega=0)=1$ $(B.3)$, the constraint $|A(\omega)-1|\leq\beta|\omega|^{\alpha}$ is satisfied, for $|A(\omega)|<\gamma$ $(B.5)$, by choosing β sufficiently large and α sufficiently close to zero.

(i)   Now

$$\left| A\left(\frac{\omega}{2^k}\right)-1 \right| \leq \beta \left| \frac{\omega^{\alpha}}{2^{k\alpha}} \right| \qquad (B.9)$$

so that

$$\prod_{k=1}^{K}\left| A\left(\frac{\omega}{2^k}\right) \right| = \exp\left\{ \sum_{k=1}^{K} \ln \left| 1+A\left(\frac{\omega}{2^k}\right)-1 \right| \right\}$$

$$\leq \exp\left(\beta \sum_{k=1}^{K} \left| \frac{\omega^{\alpha}}{2^{k\alpha}} \right| \right) \qquad (B.10)$$

The inequality in $(B.10)$ comes from the fact that if $x<y$, then $\ln(1+x)<\ln(1+y)<y$. Therefore, $(B.10)$ ensures (i).

(ii)   For any ω, there is an integer n such that

$$2^{n-1}\leq 1+|\omega|\leq 2^{n} \qquad (B.11)$$

This is because

$$n-1\leq \log_2(1+|\omega|)\leq n \qquad (B.12)$$

i.e., the log of a number always lies between two integers $n-1$ and n. But from $(B.10)$

$$\prod_{k=n}^{\infty}\left| A\left(\frac{\omega}{2^k}\right) \right| \leq \exp\left\{ \beta \sum_{k=n}^{\infty} \left| \frac{\omega^{\alpha}}{2^{k\alpha}} \right| \right\} \qquad (B.13)$$

and

$$\left| \frac{\omega}{2^k} \right| \leq 1 \text{ for all } k\geq n \qquad (B.14)$$

which is a consequence of $(B.11)$. Hence

$$\prod_{k=n}^{\infty} \left| A\left(\frac{\omega}{2^k}\right) \right| \leq c' \ , \qquad \text{independent of } \omega \qquad (B.15)$$

where c' is some constant. Next, write

$$\prod_{k=1}^{n-1} \left| A\left(\frac{\omega}{2^k}\right) \right| = \left| A\left(\frac{\omega}{2}\right) \right| \left| A\left(\frac{\omega}{2^2}\right) \right| \cdots \left| A\left(\frac{\omega}{2^{n-1}}\right) \right| \qquad (B.16)$$

Since

$$\max \left| A\left(\frac{\omega}{2^k}\right) \right| = \max |A(\omega)| \qquad (B.17)$$

applying $(B.8)$ to $(B.16)$ yields

$$\prod_{k=1}^{n-1} \left| A\left(\frac{\omega}{2^k}\right) \right| \leq B^{n-1} \qquad (B.18)$$

and via $(B.12)$ and $(B.8)$

$$\prod_{k=1}^{n-1} \left| A\left(\frac{\omega}{2^k}\right) \right| \leq B^{\log_2(1+|\omega|)} = (1+|\omega|)^b \qquad (B.19)$$

Next,

$$\left| \frac{1+z^{-1}}{2} \right|^L_{z=e^{j\omega}} = \left| \cos\left(\frac{\omega}{2}\right) \right|^L = \left| \frac{\sin(\omega)}{2\sin(\omega/2)} \right|^L \qquad (B.20)$$

so that

$$\prod_{k=1}^{\infty} \left| \frac{1+e^{-j\omega/2^k}}{2} \right|^L = \lim_{M \to \infty} \prod_{k=1}^{M} \left| \frac{\sin(\omega/2^k)}{2\sin(\omega/2^{k+1})} \right|^L \qquad (B.21)$$

In the product of terms in $(B.21)$, the denominator of the kth term always cancels the numerator of the $(k+1)$th term, hence the limit becomes

$$\lim_{M \to \infty} \left| \frac{\sin(\omega/2)}{2^M \sin(\omega/2^{M+1})} \right|^L = \left| \frac{\sin(\omega/2)}{\omega/2} \right|^L \leq 1^L \qquad (B.22)$$

From $(B.11)$,

$$1 \leq \frac{2^n}{1+|\omega|} \qquad (B.23)$$

hence

$$\left| \frac{\sin(\omega/2)}{\omega/2} \right|^L \leq \frac{c''}{(1+|\omega|)^L} \qquad (B.24)$$

where c'' is some constant. Combining $(B.15)$, $(B.19)$ and $(B.24)$ gives

$$|\Phi(\omega)| \leq \frac{c}{(1+|\omega|)^{L-b}} \quad , \quad c = c'c'' \qquad (B.25)$$

(iii) If $B < 2^{L-1}$, then substituting $(B.8)$ into $(B.25)$ yields

$$|\Phi(\omega)| \leq \frac{c}{(1+|\omega|)^r} < \frac{c}{|\omega|^r} \quad , \quad r > 1 \qquad (B.26)$$

which implies [3] that $\phi(t)$ and its derivatives of order up to $r-1$ exist and of order up to $r-2$ are continuous. Thus as long as $r > 1$, $\phi(t)$ is continuous. The number r, not necessarily integer, is termed regularity [18] of the filter $G(z)$ and measures the smoothness of $\phi(t)$. In view of (iii), PR-QMF must have regularity greater than unity in order to generate continuous $\phi(t)$ and $\psi(t)$. This in turn implies that $G(z)$ must have at least two zeros at $z = -1$. The most regular filter is one whose $L - b = $ maximum. However, simply choosing $L = $ maximum does not lead to the most regular filter [2]. When $L = p/2 = $ maximum, the filter is maximally flat, where the number of vanishing derivatives of $|G(\omega)|^2$ at $\omega = 0$ and $\omega = \pi$ are equal.

The above theorem is an adaptation from [13]. An alternate theorem is in [18], stating that if

$$\sum_k |a(k)| |k|^\varepsilon < \infty \quad , \qquad \text{for some } \varepsilon > 0 \qquad (B.27)$$

and

$$B \le 2^{L-1} \qquad (B.28)$$

then $\phi(t)$ is a continuous function. This is a sufficient but not necessary condition. Additionally, since $A(z)$ is an FIR filter with a finite number of zeros, its coefficients will always satisfy $(B.27)$ and the only check to perform is $(B.28)$. Choosing $L = p/2$ will always guarantee continuous $\phi(t)$ and $\psi(t)$.

Appendix C - Solution of a Polynomial in $\cos(\omega)$

Given the equation

$$(1-x)^L p(x) + x^L p(1-x) = 1 \qquad (C.1)$$

where

$$p(x) = |A(\omega)|^2 \quad , \quad p(1-x) = |A(\omega+\pi)|^2 \qquad (C.2)$$

and

$$x = \sin^2\left(\frac{\omega}{2}\right) = \frac{1-\cos(\omega)}{2} \qquad (C.3)$$

The problem is to find an explicit expression for $p(x)$ in terms of x.

Let

$$R(\omega) = |G(\omega)|^2 \qquad (C.4)$$

and since L is a design parameter, let $L = p/2$. This specifies $G(\omega)$ to be a maximally flat filter [22] with $R(\omega)$ having a maximum number of derivatives equal to zero at $\omega = 0$ and $\omega = \pi$ equally, i.e.

$$\frac{d^k R(\omega)}{d\omega^k} = F^{(k)}(\omega)\big|_{\omega=0,\pi} = 0 \quad , \qquad k = 1,2,\ldots,\frac{p}{2} \qquad (C.5)$$

Intuitively, vanishing of the derivatives of a function at a certain point implies flatness at that point.   The higher the number of vanishing derivatives, the higher the degree of smoothness, which improves linearly with increasing p.

From (3.168) and (C.4),

$$R(x) = (1-x)^L p(x) \qquad (C.6)$$

Direct differentiation confirms that

$$F^{(k)}(0) = 0 \quad , \qquad k = 1,2,\ldots,\frac{p}{2} \qquad (C.7)$$

With

$$p(x) = \sum_{l=0}^{L-1} p_l x^l \qquad (C.8)$$

the following finds the p_l by imposing the constraints $p(0)=1$ (B.3) and (C.7) on (C.8). Change (C.6) to

$$p(x) = R(x)(1-x)^{-L} \qquad (C.9)$$

Its lth derivative is, after some algebra

$$\left. \frac{d^l p(x)}{dx^l} \right|_{x=0} = \sum_{n=0}^{l} \binom{l}{n} F^{(n)}(0) \left[\frac{d^{l-n}(1-x)^{-L}}{dx^{l-n}} \right]_{x=0} \qquad (C.10)$$

where

$$\binom{l}{n} = \frac{l(l-1)(l-2)\ldots(l-n+1)}{n!} \qquad (C.11)$$

are the binomial coefficients. In view of (C.7), the only non-zero term in (C.10) is when $n=0$. Hence with $p(0)=1$,

$$\left. \frac{d^l p(x)}{dx^l} \right|_{x=0} = \left. \frac{d^l (1-x)^{-L}}{dx^l} \right|_{x=0}$$

$$= L(L+1)\ldots(L+l-1) \qquad (C.12)$$

But from (C.8)

$$p_l = \frac{1}{l!} \left. \frac{d^l p(x)}{dx^l} \right|_{x=0} \qquad (C.13)$$

so that

$$p_l = \binom{L+l-1}{l} \qquad (C.14)$$

or

$$|A(\omega)|^2 = \sum_{l=0}^{L-1} \binom{L+l-1}{l} \sin^{2l}\left(\frac{\omega}{2}\right) \tag{C.15}$$

By using the relationship [13]

$$\sum_{l=0}^{L-1} \binom{L+l-1}{l} = \binom{2L-1}{L-1} = \frac{1}{2}\left\{\binom{2L-1}{L-1} + \binom{2L-1}{L}\right\}$$

$$\leq \frac{1}{2} \sum_{l=0}^{2L-1} \binom{2L-1}{l} = 2^{2(L-1)} \tag{C.16}$$

it follows immediately from $(C.15)$ that $B = \max_{\omega} |A(\omega)| < 2^{L-1}$. Hence $\phi(t)$ is

continuous.

Chapter 4

CURRENT TOPICS

The subject of wavelets is developing rapidly and this Chapter attempts to cover some of the more recent research work, not in an in-depth fashion but rather in a quick tour format.

4.1 Wavelet Packets

In MRA, Figure 3.1, the input spectrum at each stage is always split into two bands at a time; the higher band becomes one of the outputs while the lower band is again split into two bands, etc. Now each splitting could be into several bands at a time. In addition, there could be a further splitting of the higher bands as well. There is a wavelet interpretation to this generalization of MRA and the outputs are called wavelet packets. In a way, MRA and dyadic grid DPWT, in providing an octave band, constant-Q analysis, are rather restrictive in the available frequency resolution. Wavelet packets, through arbitrary band splitting, can have frequency resolutions different from the octave band constant-Q scheme. An adaptive system can then choose the most suitable resolutions (combination of wavelet packets) to represent a particular signal.

The defining two-scale equations for wavelet packets are [27]

$$\psi^{(2k)}(t) = 2 \sum_{l=0}^{p-1} g(l)\, \psi^{(k)}(2t - l) \tag{4.1}$$

$$\psi^{(2k+1)}(t) = 2 \sum_{l=0}^{p-1} h(l)\, \psi^{(k)}(2t - l) \tag{4.2}$$

with initial condition

$$\psi^{(1)}(t) = \psi(t) \tag{4.3}$$

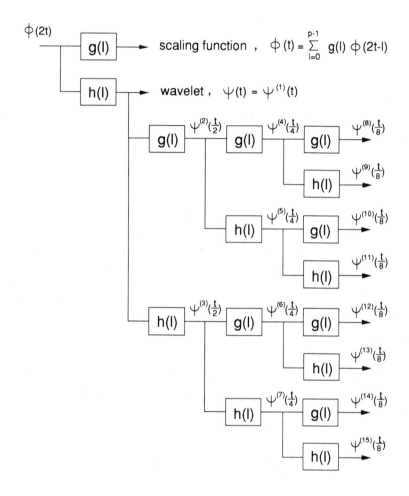

Figure 4.1 Wavelet packets generation

where $g(l)$ and $h(l)$ are the same filter coefficients as those for MRA. Wavelet packets come from a linear combination of the scaled and translated versions of a mother wavelet $\psi(t)$, as depicted in Figure 4.1. Both input and output of each block in this figure are continuous and therefore the block does not represent a digital filtering operation. Rather, it denotes a linear combination of the input in the form of (4.1) or (4.2).

By following the steps that produce the transform (3.38) from (3.19), the Fourier transform of a wavelet packet, from (4.1), is

$$\Psi^{(2^i)}(\omega) = \prod_{l=1}^{i} G\left(\frac{\omega}{2^l}\right) \Psi\left(\frac{\omega}{2^i}\right) \tag{4.4}$$

$$\Psi^{(2^i+1)}(\omega) = H\left(\frac{\omega}{2}\right) \prod_{l=2}^{i} G\left(\frac{\omega}{2^l}\right) \Psi\left(\frac{\omega}{2^i}\right) \quad , \qquad i = 1, 2, \ldots \tag{4.5}$$

Other $\Psi^{(k)}(\omega)$ follows similarly. Due to repeated band filtering, $\psi^{(k)}(t)$ for large k approaches pure tones. The wavelet packets from PR-QMF also form an orthonormal set and are quite suitable for signal decomposition [28].

4.2 Discrete Time Wavelet Transform

Section 2.1 briefly mentioned the discrete time wavelet transform (DTWT) which is a relatively new development. It is only recently that the theoretical basis for the DTWT becomes available in the literature [14,29].

With reference to Figure 3.4, the transform of the wavelet coefficients at the mth stage is (assuming ideal filters, hence ignoring the image frequencies)

$$D^m(\omega) = \sum_n d_n^m e^{-j\omega n} = \sqrt{2} \, \overline{H}\left(\frac{\omega}{2}\right) C^{m-1}\left(\frac{\omega}{2}\right) \tag{4.6}$$

where

$$\overline{H}(\omega) = \sum_l \overline{h}(l) e^{-j\omega l} \tag{4.7}$$

and

$$C^{m-1}(\omega) = \sum_n c_n^{m-1} e^{-j\omega n} \tag{4.8}$$

In (4.6), the $\omega/2$ dependency is due to decimation by two of the output of $\overline{h}(l)$. From Figure 3.1,

$$C^m(\omega) = 2^{\frac{m}{2}} \prod_{k=1}^{m} \overline{G}\left(\frac{\omega}{2^k}\right) \tilde{S}\left(\frac{\omega}{2^m}\right)$$

(4.9)

where

$$\tilde{S}(\omega) = \sum_n s(n)\, e^{-j\omega n}$$

(4.10)

Substituting (4.9) into (4.6) gives

$$D^m(\omega) = 2^{\frac{m}{2}} \overline{H}\left(\frac{\omega}{2}\right) \prod_{k=1}^{m-1} \overline{G}\left(\frac{\omega}{2^{k+1}}\right) \tilde{S}\left(\frac{\omega}{2^m}\right)$$

(4.11)

whose inverse Fourier transform is

$$d_n^m = \frac{1}{2\pi} \int \overline{F}_m(\omega)\, \tilde{S}\left(\frac{\omega}{2^m}\right) e^{j\omega n}\, d\omega$$

(4.12)

with

$$\overline{F}_m(\omega) = 2^{\frac{m}{2}} \overline{H}\left(\frac{\omega}{2}\right) \prod_{k=1}^{m-1} \overline{G}\left(\frac{\omega}{2^{k+1}}\right)$$

(4.13)

Putting (4.10) into (4.12) leads to

$$d_n^m = \frac{1}{2\pi} \int \overline{F}_m(\omega) \sum_k s(k)\, e^{-j\frac{\omega}{2^m}k}\, e^{j\omega n}\, d\omega$$

(4.14)

and

$$d_n^m = \sum_k s(k)\, \overline{f}_m(2^m n - k)$$

(4.15)

where

$$\overline{f}_m(n) = \frac{1}{2\pi} \int \overline{F}_m(\omega)\, e^{j\frac{\omega}{2^m}n}\, d\omega$$

(4.16)

Since the d_n^m are the wavelet transform coefficients, (4.15) has the interpretation that they come from a DTWT on $s(n)$, with respect to the analysis wavelet (discrete time) $\bar{f}_m(k)$. Similarly, in (4.9), let the discrete scaling function for decomposition be

$$\bar{q}_m(k) = \frac{1}{2\pi} \int 2^{\frac{m}{2}} \prod_{k=1}^{m} \bar{G}\left(\frac{\omega}{2^k}\right) e^{j\frac{\omega}{2^m}k} d\omega \tag{4.17}$$

so that

$$c_n^m = \sum_k \bar{q}_m(2^m n - k) s(k) \tag{4.18}$$

For reconstruction, refer to Figure 3.2 where

$$C^m(2^m \omega) = 2^{\frac{1}{2}} C^{m+1}(2^{m+1}\omega) G(2^m \omega)$$
$$+ 2^{\frac{1}{2}} D^{m+1}(2^{m+1}\omega) H(2^m \omega) \tag{4.19}$$

Iterating (4.19) from $m = 0$ to L, and noting that $C^0(\omega) = \bar{S}(\omega)$, gives

$$\bar{S}(\omega) = \sum_m F_m(\omega) D^m(2^m \omega) + Q_L(\omega) C^L(2^L \omega) \tag{4.20}$$

where

$$F_m(\omega) = 2^{\frac{m}{2}} H(2^{m-1}\omega) \prod_{k=0}^{m-2} G(2^k \omega) \tag{4.21}$$

and

$$Q_L(\omega) = 2^{\frac{L}{2}} \prod_{k=1}^{L} G(2^{k-1}\omega) \tag{4.22}$$

The inverse Fourier transform of (4.20) is

$$s(n) = \frac{1}{2\pi} \sum_m \int F_m(\omega) \, e^{j\omega n} \sum_k d_k^m \, e^{-j2^m \omega k} \, d\omega$$

$$+ \frac{1}{2\pi} \int Q_L(\omega) \, e^{j\omega n} \sum_k c_k^L \, e^{-j2^L \omega k} \, d\omega \qquad (4.23)$$

or

$$s(n) = \sum_m \sum_k d_k^m \frac{1}{2\pi} \int F_m(\omega) \, e^{j\omega(n-2^m k)} \, d\omega$$

$$+ \sum_k c_k^L \frac{1}{2\pi} \int Q_L(\omega) \, e^{j\omega(n-2^L k)} \, d\omega \qquad (4.24)$$

Let

$$q_L(n) = \frac{1}{2\pi} \int Q_L(\omega) \, e^{j\omega n} \, d\omega \qquad (4.25)$$

and

$$f_m(n) = \frac{1}{2\pi} \int F_m(\omega) \, e^{j\omega n} \, d\omega \qquad (4.26)$$

be the inverse Fourier transforms of $Q_L(\omega)$ and $F_m(\omega)$. Then the inverse DTWT is

$$s(n) = \sum_{m=1}^{L} \sum_k d_k^m f_m(n - 2^m k) + \sum_k c_k^L q_L(n - 2^L k) \qquad (4.27)$$

with $f_m(n)$ the discrete synthesis scaling function and $q_L(n)$ the discrete synthesis wavelet function.

Thus MRA, which already has the dual roles of (i) subband decomposition and (ii) DPWT of $s(t)$, can take on simultaneously yet another of (iii) DTWT of $s(n)$. The immediate question is: what is the relationship between the discrete time $\bar{f}_m(k)$, $\bar{q}_m(k)$, $f_m(k)$ and $q_m(k)$ and the continuous time $\phi_{mn}(t)$ and $\psi_{mn}(t)$? In general, they have no explicit connections.   However, under certain conditions, the DTWT is an exact discrete time implementation of the DPWT [14], as the next paragraph shows.

From (2.9) with $a_o = 2$ and $\tau_o = 1$,

$$DPWT(m,n) = 2^{-\frac{m}{2}} \int s(t)\, \psi(2^{-m}t - n)\, dt \qquad (4.28)$$

The discretization of this integral is

$$DTWT(m,n) = 2^{-\frac{m}{2}} \sum_k s(k)\, \psi(2^{-m}k - n) \qquad (4.29)$$

Consider the decomposition scheme of Figure 4.2, which is similar to MRA except that there is no decimation in the output of the lower branches.   The problem is to specify conditions on the filters $a(l)$ and $b(l)$ so that the scheme implements (4.29) and to determine the $\psi(t)$ associated with these filters.   The output of the first branch is

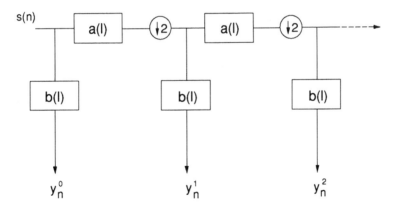

Figure 4.2   Discrete time wavelet transform implementation

$$y_n^0 = \sum_{k=n-x}^{n+x} s(k)\, b(n-k) \qquad (4.30)$$

For notational convenience which becomes clear later, the index for the filter coefficients $b(l)$ runs from $-x = -(p-1)/2$ to x and the support for $\psi(n)$ is also from $-x$ to x. Comparing (4.30) with (4.29), which is

$$\text{DTWT}(0,n) = \sum_{k=n-x}^{n+x} s(k)\,\psi(k-n) \tag{4.31}$$

shows that $y_n^0 = \text{DTWT}(0,n)$ if

$$\psi(l) = b(-l) \tag{4.32}$$

Next, the Fourier transform of y_n^1 is

$$Y^1(\omega) = \tilde{S}\!\left(\frac{\omega}{2}\right) A\!\left(\frac{\omega}{2}\right) B(\omega) \tag{4.33}$$

where $\tilde{S}(\omega/2)$ is (4.37), $A(\omega)$ and $B(\omega)$ are respectively the Fourier transform of $a(l)$ and $b(l)$, and the $\omega/2$ dependency is a result of downsampling by two the output of $a(l)$. The Fourier transform of (4.29) is, for $m = 1$,

$$D^1(\omega) = \frac{1}{\sqrt{2}} \sum_n \sum_{k=2n-2x}^{2n+2x} s(k)\,\psi\!\left(\frac{k}{2}-n\right) e^{-j\omega n} \tag{4.34}$$

which becomes, on letting $k = 2n + l$,

$$D^1(\omega) = \frac{1}{\sqrt{2}}\,\hat{\Psi}^*(\omega)\,\tilde{S}\!\left(\frac{\omega}{2}\right) \tag{4.35}$$

where

$$\hat{\Psi}(\omega) = \sum_{l=-2x}^{2x} \psi\!\left(\frac{l}{2}\right) e^{-j\omega\frac{l}{2}} \tag{4.36}$$

and

$$\tilde{S}\!\left(\frac{\omega}{2}\right) = \sum_n s(2n)\, e^{-j\omega n} \tag{4.37}$$

Note that because of (4.32),

$$B(\omega) = \sum_{l=-x}^{x} b(l) e^{-j\omega l} = \sum_{l=-x}^{x} \psi(-l) e^{-j\omega l} = \Psi^*(\omega) \qquad (4.38)$$

However, $\Psi^*(\omega) \neq \hat{\Psi}^*(\omega)$ of (4.36) because the latter is dependent on values of $\psi(l/2)$, l odd, which do not appear in (4.38). Indeed, if y_n^1 is to be equal to DTWT$(1,n)$, the $\psi(l/2)$, l odd, need to be specified in terms of $a(l)$ and $b(l)$. To this end, equating (4.33) to (4.35) and using (4.38) gives

$$A\left(\frac{\omega}{2}\right)\Psi^*(\omega) = \frac{1}{\sqrt{2}}\hat{\Psi}^*(\omega) \qquad (4.39)$$

or

$$\sum_{k=-2x}^{2x} a(k) e^{-j\frac{\omega}{2}k} \Psi^*(\omega) = \frac{1}{\sqrt{2}}\hat{\Psi}^*(\omega) \qquad (4.40)$$

where $a(l)$ is taken to be a $2p$-1 odd order filter. Multiplying both sides by $\exp(-j\omega l/2)/2\pi$ and integrating over ω yields

$$\sum_{k=-2x}^{2x} a(k) \psi\left(\frac{k+l}{2}\right) = \frac{1}{\sqrt{2}}\psi\left(\frac{l}{2}\right) , \qquad l = -2x, -2x+1, \ldots, 2x \qquad (4.41)$$

Let

$$a(l) = \begin{cases} \dfrac{1}{\sqrt{2}} , & l = 0 \\ 0 , & l \neq 0, \; even \\ \text{to be specified} , & l \; odd \end{cases} \qquad (4.42)$$

Then for l even, $\psi(l/2)$ is identical to $b(-l/2)$, as specified in (4.32). For l odd, $\psi(l/2)$ is a weighted sum of the $\psi(i)$, i = integer, i.e., the filter $a(k)$ interpolates the $\psi(i)$ to give $\psi(l/2)$, l odd. Hence $y_n^1 = $ DTWT$(1,n)$ if the samples $\psi(k/2-n)$, k odd is given by (4.41).

A filter that satisfies (4.42) is à trous (with holes) [14]. It is a lowpass interpolation filter. Following similar step as above will produce interpolation for other values of m of $\psi(k/2^m-n)$. The conclusion is that Figure 4.2 is an implementation of the DTWT of (4.29) if $a(l)$ is à trous and the wavelet samples are given by (4.32) when l = integer, and the other samples $\psi(k/2^m-n)$ come from successive interpolation of the $\psi(l)$ via à trous filters. Note that to allow for decimation by two, $a(l)$, besides being lowpass, should reduce the bandwidth by two as well. For an example of an à trous filter, consider the coefficients

$$a(l)=\frac{2}{\sqrt{2}}\sum_{m=0}^{p-1}g(m)g(m+l) \quad , \quad l=-(p-1),\ldots,0,\ldots,p-1 \quad (4.43)$$

where $g(m)$ are the coefficients of the lowpass PR-QMF of (3.181). Using (3.112) and by direct computation give the values of $a(l)$ as $-1/(16\sqrt{2}),0,9/(16\sqrt{2}),1/\sqrt{2},9/(16\sqrt{2}),0,-1/(16\sqrt{2})$. They satisfy (4.42) and $a(l)$ is à trous.

4.3 Signal Processing Applications

Due to the time-frequency localization capability of the wavelet transform, it obviously has similar applications as the STFT and some examples are in Section 1.5. The scalogram is an alternative to the spectrogram for functions such as feature extraction, parameter estimation and pattern recognition of nonstationary signals.

The scalogram also exhibits distinctive patterns at the instant of an abrupt change in a signal; [30] has shown how this may be useful in studying the characteristics of speech and musical sounds, while [31] applies it to estimate the pitch period in speech. Intuitively, abrupt changes in a signal invariably produce high frequency components. Thus at small scales a, which correspond to high frequency regions, the scalogram will have high magnitudes. More specifically, consider the wavelet transform of a delta function occuring at $t=t_o$

$$\text{CWT}(a,\tau)=\frac{1}{\sqrt{a}}\int\psi\left(\frac{t-\tau}{a}\right)\delta(t-t_o)\,dt \quad (4.44)$$

$$=\frac{1}{\sqrt{a}}\psi\left(\frac{t_o-\tau}{a}\right) \quad (4.45)$$

As $a \to 0$, $|\mathrm{CWT}(a,\tau)| \to \delta(t_o - \tau)$. At small a's, $|\mathrm{CWT}(a,\tau)|$ has large magnitudes at $\tau = t_o$, the instant of occurrence of the impulse. If the wavelet is complex, the wavelet transform phase displays an additional feature at the instant of an abrupt change. Let $\psi(t)$ be the wavelet of (2.4), then from (4.45)

$$\mathrm{CWT}(a,\tau) = \frac{1}{\sqrt{a}} \psi\left(\frac{t_o - \tau}{a}\right) = \frac{1}{\sqrt{a}} e^{-j\omega_o\left(\frac{t_o-\tau}{a}\right)} e^{-\left(\frac{t_o-\tau}{a}\right)^2} \tag{4.46}$$

The phase of $\mathrm{CWT}(a,\tau)$ is

$$\theta = -\omega_o\left(\frac{t_o - \tau}{a}\right) \tag{4.47}$$

or

$$a = \frac{\omega_o \tau}{\theta} - \frac{\omega_o t_o}{\theta} \tag{4.48}$$

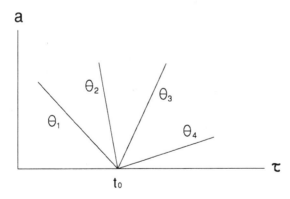

Figure 4.3 Constant phase plot of the wavelet transform of an impulse

A plot of a versus τ for constant θ's is in Figure 4.3 (see also Figure 2.8b). It shows that constant phase lines converge towards the point of discontinuity on the a-τ plane. When a discontinuity is not an impulse, such distinctive features of course will not appear. Nevertheless, patterns close to the ideal may be sufficient to indicate a discontinuity. The accuracy of pitch estimation is, however, dependent on the particular wavelet used [32].

The wavelet transform is an inner product operation that measures the similarity or cross-correlation between a signal $s(t)$ and $\psi((t-\tau)/a)$ at the scale a and shift τ. Thus

$$\text{CWT}(a,\tau)=\frac{1}{\sqrt{a}}\int s(t)\,\psi\!\left(\frac{t-\tau}{a}\right)dt \tag{4.49}$$

In active sonar or radar, a modulated (by pure tone or linear FM) pulse $s(t)$ impinges on a moving target resulting in an echo $\hat{s}(t)$ which is a Doppler shifted and delayed version of $s(t)$. If $s(t)$ is narrowband, i.e., the modulation is by a pure tone, then the detection process consists of finding τ and ω_D to maximize the narrowband cross-ambiguity function

$$\text{NAF}=\int s(t)\,\hat{s}(t-\tau)\,e^{-j\omega_D t}\,dt \tag{4.50}$$

The parameter τ gives the delay of $\hat{s}(t)$ with respect to $s(t)$ and determines the target range. The Doppler frequency is ω_D, which is the frequency shift of $\hat{s}(t)$ from $s(t)$ and determines the target velocity with respect to the transmitter. When the bandwidth of $s(t)$ is large, usually this means that the bandwidth to center frequency ratio is greater than unity, the NAF is no longer valid. This is because the modulation rate is rapid enough to cause phase variation in the carrier; $\hat{s}(t)$ now becomes a compressed (or stretched, by the moving target), and delayed version of $s(t)$. For detection, it is necessary to use the wideband cross-ambiguity function

$$\text{WAF}=\int s(t)\,\hat{s}\!\left(\frac{t-\tau}{a}\right)dt \tag{4.51}$$

The computation of the NAF is by STFT (1.66) and the WAF is by CWT (4.49). Taking $\hat{s}(t)$ as the wavelet, the wavelet transform of $s(t)$ determines the shape of the WAF. For optimum detection, [33] discussed the design of a $s(t)$ whose wavelet transform has a narrow and sharp peak. Note that in both radar and sonar, the time scaling a in the echo is very close, but different, from unity, when the target is in motion.

Finally, the most important application for wavelet is probably in signal compression or coding, for image or speech. Signal coding by wavelet packets appears particularly promising at present [27].

REFERENCES

1. Lim, J.S., *Two-dimensional Signal and Image Processing*, Prentice Hall, Englewood Cliffs, NJ, 1990.
2. Akansu, A.N., and R.A. Haddad, *Multiresolution Signal Decomposition*, Academic Press, Inc., San Diego, CA 1992.
3. Papoulis, A., *Signal Analysis*, McGraw-Hill, New York, NY 1977.
4. Daubechies, I., "The Wavelet Transform, Time-Frequency Localization and Signal Analysis", *IEEE Trans. Inform. Theory*, Vol. 36, No. 5, pp. 961-1005, Sept. 1990.
5. Zadeh, L.A., and C.A. Desoer, *Linear System Theory: The State Space Approach*, McGraw-Hill, New York, NY, 1963.
6. Gantmakher, F.R., *Theory of Matrices,* Vols. 1 and 2, Chelsea Publishing Co., New York, NY, 1959.
7. Jenkins, G.M., and D.G. Watts, *Spectral Analysis and Its Applications,* Holden-Day, San Francisco, CA, 1968.
8. Oppenheim, A.V., and R.W. Schafer, *Discrete-Time Signal Processing*, Prentice Hall, Inc., Englewood Cliffs, NJ, 1989.
9. Cohen, L., "Time-Frequency Distributions - A Review", *Proc. IEEE*, Vol. 77, No. 7, pp. 941-981, July 1989.
10. Lim, J.S., and A.V. Oppenheim, Editors, *Advanced Topics in Signal Processing,* Prentice Hall, Englewood Cliffs, NJ, 1988.
11. Oppenheim, A.V., Editor, *Applications of Digital Signal Processing,* Prentice Hall, Englewood Cliffs, NJ, 1978.
12. Haykin, S., Editor, *Advances in Spectrum Analysis and Array Processing*, Vol. I, Prentice Hall, Englewood Cliffs, NJ, 1991.
13. Chui, C.K., *An Introduction to Wavelets*, Academic Press, San Diego, CA 1992.
14. Shensa, M.J., "The Discrete Wavelet Transform: Wedding the À' Trous and Mallat Algorithms", *IEEE Trans. Signal Processing*, Vol. 40, No. 10, pp. 2464-2482, Oct., 1992.
15. Rioul, O., and M. Vetterli, "Wavelets and Signal Processing", *IEEE SP Magazine*, pp. 14-38, Oct. 1991.
16. Flanagan, J.L., *Speech Analysis, Synthesis, and Perception*, 2nd Ed., Springer-Verlag, New York, NY, 1972.

17. Nielsen, R.O., *Sonar Signal Processing*, Artech House, Boston, MA 1991.

18. Daubechies, I., "Orthonormal Basis of Compactly Supported Wavelets", *Comm. in Pure and Applied Math.*, Vol. 41, No.7, pp. 909-996, 1988.

19. Mallat, S.G., A Theory for Multiresolution Signal Decomposition: The Wavelet Representation", *IEEE Trans. Pattern Analysis, Machine Intelligence*, Vol. 11, No. 7, pp. 674-693, July 1989.

20. Tretter, S.A., *Introduction to Discrete-Time Signal Processing*, John Wiley and Sons, New York, NY, 1976.

21. Cody, M.A., "The Fast Wavelet Transform", *Dr. Dobb's Journal*, pp. 16-91, Apr. 1992.

22. Vaidyanathan, P.P., *Multirate Systems and Filter Bands*, Prentice Hall, Englewood Cliffs, NJ, 1993.

23. Herley, C., and M. Vetterli, "Wavelets and Recursive Filter Banks", *IEEE Trans. Signal Processing*, Vol. 41, No. 8, pp. 2536-2556, Aug. 1993.

24. Vetterli, M., and C. Herley, "Wavelets and Filter Banks: Theory and Design", *IEEE Trans. Signal Processing*, Vol. 40, No. 9, pp. 2207-2232, Sept. 1992.

25. Strang, G., "Wavelets and Dilation Equations: A Brief Introduction", *Society for Industrial and Applied Mathematics Review*, Vol. 31, No. 4, pp. 614-627, Dec. 1989.

26. Karlsson, G., and M. Vetterli, "Theory of Two-Dimensional Multirate Filter Banks", *IEEE Trans. Signal Processing*, Vol. 38, No. 6, pp. 925-937, June 1990.

27. Ruskai, M.B., et. al., Editors, *Wavelets and Their Applications*, Jones and Bartlett Publishers, Inc., Boston, MA, 1992.

28. Chui, C.K., Editor, *Wavelets: A Tutorial in Theory and Applications*, Academic Press, Inc., Boston, MA, 1992.

29. Rioul, O., "A Discrete-Time Multiresolution Theory", *IEEE Trans. Signal Processing*, Vol. 41, No. 8, pp. 2591-2606, Aug. 1993.

30. Kronland-Martinet, R., J. Morlet and A. Grossman, "Analysis of Sound Patterns through Wavelet Transforms", *Intl. J. Pattern Recognition and Artificial Intelligence*, Vol. 1, No. 2, pp. 273-302, 1987.

31. Kadambe, A., and G.F. Boudreaux-Bortels, "Application of Wavelet Transform for Pitch Detection of Speech Signals", *IEEE Trans. Inform. Theory*, Vol. 38, No. 2, pp. 917-924, Mar. 1992.

32. Kadambe, S., and G.F. Boudreaux-Bartels, "A Comparison of Wavelet Functions for Pitch Detection of Speech Signals", *Proc. ICASSP-91*, pp. 449-452, Toronto, 1991.

33. Jin, Q., K.M. Wong and Z.Q. Luo, "Wideband Time Delay and Doppler Stretch Estimation: The Application of Wavelet Transform and the Optimum Signal", *Proc. ICASSP-93*, pp. I241-I244, Minneapolis, 1993.

34. Abry, P. and P. Flandrin, "On the Initialization of the Discrete Wavelet Transform Algorithm", *IEEE Signal Processing Letters*, Vol. 1, No. 2, pp. 32-34, Feb. 1994.

35. Xia, X.G., C.C.J. Kuo and Z. Zhang,"Wavelet Coefficient Computation with Optimal Prefiltering", *IEEE Trans. Signal Processing*, Vol. 42, No. 8, pp. 2191-2197, Aug. 1994.

Books on, or contain material on, Wavelets

1. *Wavelets and Their Applications*, Editor: M.B. Ruskai, Jones and Bartlett Publishers, Boston, MA, 1992.
2. *Wavelets: Mathematics and Applications*, Editors: J.J. Benedetto and M.V. Frazier, CRC Press, Boca Raton, Florida, 1994.
3. *Wavelet Theory and Its Applications*, R.K. Young, Kluwer Academic Press, Boston, MA, 1993.
4. *An Introduction to Wavelets*, C.K. Chui, Academic Press, Inc., Boston, MA 1992.
5. *Wavelets: A Tutorial in Theory and Applications*, Editor: C.K. Chui, Academic Press, Inc., Boston, MA, 1992.
6. *Advances in Spectrum Analysis and Array Processing, Vol. I*, Editor: S. Haykin, Prentice Hall, Englewood Cliffs, NJ, 1991.
7. *Multirate Systems and Filter Banks*, P.P. Vaidyanathan, Prentice Hall, Englewood Cliffs, NJ, 1993.
8. *Multiresolution Signal Decomposition*, A.N. Akansu and R.A. Haddad, Academic Press, Inc., San Diego, CA 1992.
9. *Ten Lectures on Wavelets*, I. Daubechies, CBMS-NSF (Conference Board of the Mathematical Science-National Science Foundation) series in Applied Mathematics, Society for Industrial and Applied Mathematics, Philadelphia, PA, 1992.

E-mail address for Wavelet Digest:

Submissions for wavelet digest:
wavelet@math.scarolina.edu with "submit" as subject

Subscriptions for wavelet digest:
wavelet@math.scarolina.edu with "subscribe" as subject

Archive Site:
Anonymous ftp to maxwell.math.scarolina.edu, directory /pub/wavelet

INDEX